Harnessing
THE WIND
for Home Energy

Harnessing
THE WIND
for Home Energy

Dermot McGuigan

GARDEN WAY PUBLISHING
CHARLOTTE, VERMONT 05445

Printed in the United States
Second printing, August 1978

Library of Congress Cataloging in Publication Data

McGuigan, Dermot, 1949–
    Harnessing the wind for home energy.

    Bibliography: p.
    Includes index.
    1. Electric power production.   2. Wind power.
I. Title.
TK1541.M3   621.312′136   77-17916
ISBN 0-88266-118-3
ISBN 0-88266-117-5 pbk.

CONTENTS

DIRECTORIES

INTRODUCTION

I feel particularly happy about what is going on in the world today, even though it is true that we are rapidly running out of oil, upon which our whole economic structure depends. There is something perhaps more important quietly developing behind the scenes. One aspect of this new development is the urge to work closely with nature — to tap the abundance of ambient energy, whether it be solar, water, wood, or wind power.

It is becoming increasingly obvious that, while these new sources of power cannot provide so much concentrated energy as oil, they certainly can fulfill all of our needs. Various sources of natural energy are sometimes combined: For a domestic house, for example, wind power can light the home and power electric appliances while solar energy or wood-burning stoves can heat both water and space. Or, as we will show later in this book, it is possible to power the whole house directly from the wind.

There are basically two ways to approach wind power. The first, called a *low-energy system,* is to use a small wind generator (with a capacity of 200 to 1,000 watts) to charge batteries, which in turn provide power for lighting and other small electrical appliances. Most such systems are efficient, "energy conscious," and express the idea that electricity, especially the refined public utility type, is too valuable and costly to use for space and water heating — at least not when these requirements can be served by other sources of power. Low-energy systems can be quite inexpensive to set up.

The second main wind power use is to install a *large wind generator* to power all the usual domestic appliances or to supply space and water heating. Full appliance application invariably needs a large bank of batteries and an inverter, both of which can cost as much as the wind generator itself. The exception to this is to connect a device called a *Gemini inverter* between the wind generator and a nearby public utility line. If done, the Gemini will convert the varying voltage and frequency from the wind generator to be the same as the power line. Power can be stored in and drawn from the power line when required, thus avoiding the need for batteries.

Electricity needed for space and water heating is never taken through the battery or inverter from the power line, but instead is taken in a crude form direct from the wind generator. Such a system tends to be expensive to set up, though once installed there are no further costs for many years. Remember here that the average oil, gas, and electricity prices have tripled in the past few years and are likely to continue to increase.

While fuel inflation is not a sufficient reason for investing in a a wind plant now, there are other reasons for the recent resurgence of interest in windmills. It is a nice feeling to know that your source of power is in your own hands and not in those of the Arab magnates or some vast impersonal monopoly. Apart from being beautiful in themselves, windmills use a renewable and non-polluting source of power. They are fun to work with and to watch in operation.

I have yet to meet anyone who, upon first meeting with a wind generator, has not come to a standstill and stared with awe. Then, after the initial surprise, comes the inevitable flood of questions. I hope this book will answer many of your questions on wind power. Happy reading.

I am particularly grateful to the following persons, without whose assistance this book would have been impossible: Thomas Cabot, Robin Clarke, Peter Eaton, Bob Fletcher, Jack Park, Ed Salter, and Kuno Tichatschek.

DERMOT MCGUIGAN

ABOUT THE AUTHOR

Dermot McGuigan, who is of Irish origin, became interested in the alternate energy field several years ago as part owner of the manufacturing and research firm that pioneered alternate energy work in the United Kingdom. He worked full time with the company until 1976.

Previous to this, McGuigan was for four years associated with the Ecology Bookshop in London, leaving to involve himself more practically with working energy systems.

Though not a graduate civil, mechanical, or electrical engineer, McGuigan had become deeply versed in all three disciplines. He has considerable experience with windmills and wind power installations.

McGuigan's primary interest is to work with nature in the development of "alternatives" in agriculture, education, architecture, and especially energy. Born in 1949, he enjoys hang-gliding and surfing.

European editions of McGuigan's *Water Power* and *Wind Power* books are being published in England.

PLANNING
YOUR ENERGY NEEDS

Determining energy requirements is a most interesting and productive exercise, for it involves a close examination of lifestyle and priorities. What is needed and what is merely a useless burden?

Wind energy *can* power the modern all-electric suburban home, but at a very high cost. Assessing the suitability of ambient energy sources for home energy and then adapting the house to these new forms of energy is, to me at any rate, more a functional art form than a task. (I am assuming here that many of those reading this book live, or intend to live, in isolated or rural areas.)

By *ambient* energy I mean the abundance of natural energy that surrounds the home. Even in the desert there is this abundance — solar energy. Solar panels can be used for water heating for at least half of the year. A solar space-heated home is a practicality; it is also possible to retrofit an existing house to suit the sun. Combinations of solar and wind energy are compatible. During prolonged winter cloud cover, even the best insulated solar heated home will begin to feel the chill. But it is in the winter when the wind blows most.

A wind generator may be installed which will supply the normal demand for electricity throughout the year and also contribute the excess of power generated in winter to heating. The University of Massachusetts has done a lot of practical work on this combination.

In areas where timber is plentiful, the blending of woodburning energy and wind power is a natural choice. Woodburning will give space and water heating, and while cooking with it is not as easy as with gas or electricity, to those who live an unhurried life it presents no problems.

1

Again, a combination of wood and solar energy to provide water heating may be appropriate — solar in summer and wood in winter. Water power too may be available, and if so, the costs of a small turbine and a wind generator should be compared.

But whatever the combination of ambient energy sources, there always will be an essential place for wind-generated electricity. Wind power can light the home, operate radio, television and stereo, pump water, heat space and water, and perform a myriad of other useful functions.

The cost of a wind energy system is directly related to its power output, which in turn should be matched to the power need. Table 1 lists a selection of appliances and indicates the power each uses. The table is only an approximation, and each potential wind power user should complete a similar chart. This is easily done by noting the power consumed by all the appliances used and multiplying the power in watts by the time each is used.

Domestic "power guzzlers" obviously need a large and expensive wind system to fulfill their requirements, while those with more modest needs will find a small and reasonably priced wind energy system suitable. Asceticism or any other form of restriction has no place in wind power planning. But working with the wind does tend to increase energy consciousness and lead to an understanding that doing more with less is ennobling. Surely, in summer a solar clothes dryer, a clothes line, is better than a steaming, odorous electric dryer. And is not a towel a quieter and more pleasant way of drying the hair than with a noisy electric heater? You get the idea. Take a close look at what you need or want and at what you don't. Is a vacuum cleaner really any better than a broom or carpet sweeper? Are electric toothbrushes and shavers just a fad, fostered by the commercials, or do they make any real contribution to your life?

Another example is the refrigerator. In winter it becomes a heater in that it is warmer inside the box than it is outside the house. If an "outside cooler" is used there is not even a need to go outside to get the food, at least not if a well-insulated hatch is provided opening into the "cooler." I will never forget the first (and last) garbage disposal unit I used. Suffice to say that I have never since come across a more pointless and costly gadget. Now, please don't get the idea that I am preaching a life of miserable deprivation — quite the opposite — for I believe in abundance. But I feel that true

TABLE 1

Household Appliances:
Their Energy Use and Its Cost, 1977

The table below lists the amounts of electricity used in a month by many large and small household appliances. Also shown is the cost for this use as charged by a typical utility in mid-1977. The cost-per-month figures given in the right-hand columns are based only on the kilowatt hours shown in the table; normal use can vary widely depending on your locale.

Appliance	Average Wattage	Hours used per month	Percent of use time when current flows	Approximate kilowatt hours per month	Cost per month — Summer	Cost per month — Winter
air conditioner	860	150	100	130	$3.00	—
auto headbolt heater	300	300	100	90	—	$4.95
baby food warmer	165	11	100	1.8	.04	.10
blanket	150	166	50	12.45	—	.68
blender	386	.5	100	.19	.005	.01
can opener	100	.25	100	.025	.0005	.0014
clock	2	720	100	1.4	.03	.07
clothes dryer	4,856	17	100	83	1.91	4.57
coffee maker						
brew cycle	600	12.5	100	7.5	.17	.41
warm cycle	80	50	100	4.0	.09	.22
Corn popper	575	1.25	100	.72	.02	.04
cooker/fryer	1,200	3	54	2	.05	.11
dehumidifier	257	720	35	64	1.47	—
dishwasher³	1,200	30	—	30	.69	1.65
disposer	440	1	100	.5	.01	.03
egg cooker	550	2	100	1.1	.03	.06
fan, window	200	300	100	60	1.38	—
floor polisher	305	2	100	.61	.01	.03
fondue/chafing dish	800	2	46	.74	.02	.04
fry pan	1,200	11.25	62	8.37	.19	.46
Freezer (14 cubic ft.)						
manual defrost	341	720	40	98	2.25	5.39
automatic	440	720	46	146	3.36	8.03
griddle	1,200	4	76	3.65	.08	.20
hair clipper	10	3	100	.03	.0007	.002
hair dryer, soft bonnet	400	6.25	100	2.5	.06	.14

(Continued next page)

TABLE 1
(Continued)

Appliance	Average Wattage	Hours used per month	Percent of use time when current flows	Approximate kilowatt hours per month	Cost per month — Summer	Cost per month — Winter
hair dryer, hard bonnet	900	4.25	100	3.83	.09	.21
hair dryer, hand held	600	3.5	100	2	.05	.11
hair setter/curlers	350	3.25	100	1.14	.03	.06
heating pad	60	12	54	.4	.01	.02
Heating system						
burner motor	266	720	31	60	1.38	3.30
hot air fan	292	720	38	80	1.84	4.40
hot water circulator	120	720	34	30	.69	1.65
heating tape (30 ft.)	180	720	100	130	—	7.15
humidifier	177	720	20	26	—	1.40
ice cream freezer	130	1	100	.13	.003	.007
ice crusher	100	.5	100	.05	.001	.003
iron	1,100	10	52	5.72	.13	.31
juicer	90	.5	100	.045	.001	.002
knife	95	.5	100	.048	.001	.003
lighted mirror	20	9	100	.18	.004	.01
microwave oven	1,450	8	100	11.6	.27	.64
mixer, hand	80	1	100	.08	.002	.004
mixer, stand	150	1	100	.15	.003	.008
radio	71	100	100	7	.16	.39
radio/record player	100	100	100	10	.23	.55
Range[1]	12,000	8	100	98	2.25	5.39
small surface unit (on high)	1,300	8	100	10.4	.24	.58
large surface unit (on high)	2,400	8	100	19.2	.46	1.06
oven	2,660	8	26	39	.90	2.14
self cleaning process[2]	2,500	3	100	7.5	.17	.41
Refrigerator/Freezer (15 cubic ft.)						
automatic defrost	440	720	46	146	3.36	8.03
manual defrost	325	720	40	94	2.16	5.17
roaster	1,425	6	58	5	.12	.28
rotisserie	1,400	4	100	5.6	.13	.31
sewing machine	75	6	100	.45	.01	.03
shaver	15	2.5	100	.04	.001	.002
shaving cream dispenser	60	1	100	.06	.001	.002
slow cooker	200	50	100	10	.23	.55
sun lamp	279	4	100	1	.02	.06
Television, black and white						
tube type	160	182	100	29	.67	1.60
solid state	55	182	100	10	.23	.55

TABLE 1
(Continued)

Appliance	Average Wattage	Hours used per month	Percent of use time when current flows	Approximate kilowatt hours per month	Cost per month — Summer	Cost per month — Winter
Television, color						
tube type	300	183	100	55	1.27	3.03
solid state	200	183	100	37	.85	2.04
Toaster	1,100	3	100	3.3	.08	.18
Toaster oven						
toasting	1,500	2	100	3	.07	.17
oven	1,500	12	26	4.7	.11	.26
toothbrush	1.1	720	100	.8	.02	.04
vacuum cleaner	630	4	100	2.5	.06	.14
waffle iron/sandwich grill	1,200	2	80	1.9	.04	.10
warming tray	140	4	100	.56	.01	.03
Washing Machine						
automatic	512	17	100	9	.21	.50
non-automatic	286	22	100	6	.14	.33
water heater — (regular) (80 gallon tank)	3,000	720	18.7	405	9.32	22.28
water heater — off peak (80 gallon tank)	3,000	—	—	400	9.55	9.55
water heater — quick recovery	4,500	720	12.5	405	9.32	22.28
water pump	460	720	6	20	.46	1.10
Lighting (5-7 rooms)	varied	varied	varied	80-150	1.84-3.45	4.40-8.25

Notes:

1. Estimated for all range and oven units for average family for 1 month.
2. Estimated for one 3-hour cleaning per month.
3. Heating of water included under water heater.

Sources: Edison Electric Institute and American Home Appliance Manufacturers)

---

abundance is to be found in quality and self-determination, and not in heavily promoted gadgets.

By rationalizing energy demand and reducing waste, a considerable saving can be effected in purchasing a compact and simple wind energy system. The energy bonanza, as President Carter has pointed out, is due to end about 1982 when there will be a world shortage of oil. We have the opportunity to choose energy independence now and avoid the brutal impact of such future circumstances. One of the happy consequences of running out of fossil fuels is that it will bring us forward (and *not* back) to the land, to a closer and more understanding coexistence with nature, to living with the wind and the sun.

Site Selection

The one essential ingredient for the operation of a wind-driven generator, of course, is wind, and the more of it the better. Those who live in an exposed area — on a hillside, by the sea or on an open plain — are generally aware of the force of the wind, and many of them rightly feel that the wind could be put to good use.

On the other hand there are those whose homes lie shrouded by tall trees at the bottom of quiet valleys where the wind rarely blows, and when it does blow it tends to be a turbulent wind which would cause havoc with a windmill. Unfortunately, those who have the peace of the valley are unlikely to benefit from the wind.

The site selected has a bearing on two important factors: (1) possible stresses on the windmill as a result of air turbulence, and (2) the energy output of the windmill.

The most likely cause of air turbulence is local obstructions such as trees, buildings and hills. When a smooth airstream encounters a nearby obstruction, the reaction tends to be a severe buffeting, which causes sharp variations in the stresses on different parts of the generator's propeller and tower.

To avoid these destructive effects, it is essential to erect the wind generator at least 20 feet above any obstruction that lies within a 300-foot radius — preferably 30 and 500 feet respectively. The drawings in Figure 1 give a good indication of where and where not to site a windmill. Good and bad wind sites should be fairly obvious, judging from the lie of the land. In case of difficulty, tie streamers to the tallest portable pole you can get. Place the pole at different sites and watch for the rippling and curling effect of air turbulence. Turbulent air tends to roll in circles as a ball. The best site for the windmill is where the streamer holds steady.

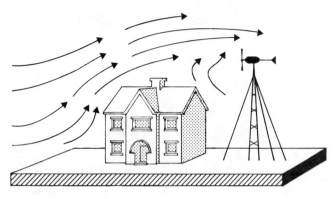

Figure 1A. The house is diverting the windstream above the generator, which should either be raised or moved.

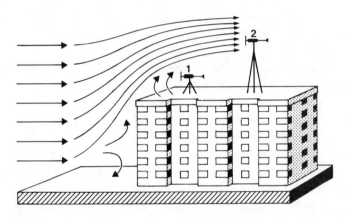

Figure 1B. Again, the building is diverting the wind stream from generator 1, but actually increases the windspeed for generator 2.

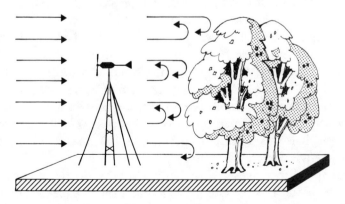

Figure 1C. Obstructions cause wind turbulence whether they are in front of or behind the generator, as here. The mast should be extended 20–30 feet above the trees or sited 300–500 feet away from them.

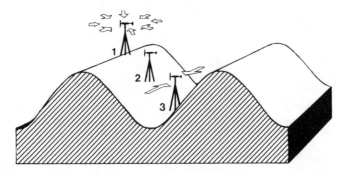

Figure 1D. Boundary considerations may rule out the best sites in hilly country. #1 is the ideal; it benefits from all wind directions. #2 is not recommended. #3 is a good site for wind coming from two directions only.

It is also necessary to site the windmill where it is well exposed to the main prevailing winds. This is important in those areas where most of the time the wind blows from only one of two directions. If you are not familiar with the area and its prevailing winds, the National Climatic Center (Federal Building, Asheville, N.C. 28801) can supply wind direction data for any part of the country.

It is generally reckoned that where the average annual wind speed is under 8 to 10 miles per hour (mph), that harnessing wind power is not a very attractive proposition. The power in such winds is very low, and it would require an enormous and expensive rotor to extract any appreciable quantity of energy.

One of the laws of wind power is that power in the wind is pro-portional to the *cube* of the wind speed. This means that a 12 mph wind has *eight* times the power of a 6 mph wind and a 24 mph wind has *sixty-four* times the power of a 6 mph wind! This is a very important characteristic, for once the average wind speed goes beyond the 10 mph mark — to say 12 mph or more — then wind power really begins to look like an exciting possibility. To put it another way, there is 73 percent more energy in a 12 mph wind than at 10 mph. It is this additional percentage that makes all the dif-ference in a wind energy system.

Wind speed increases greatly with altitude. So the higher the tower, the better it is for energy gain and a turbulence-free opera-tion. Figure 2 shows both the wind speed and the energy that can be gained by increasing tower height: for example, there is about 3.5 times the power available to a windmill at 40 feet, than at 5 feet,

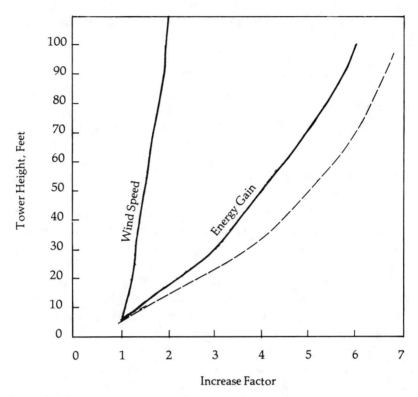

Figure 2. The increase in windspeed and energy gained by increasing tower height. Unbroken line is for flat land or sea; broken line is for rough terrain.

even though the wind speed is only about 1.5 times as great. It is true to say that *the least expensive way to get more power from the wind is to increase the height of the tower.* This tends to hold true, cost-wise, up to about 100 feet, though there are many who prefer lower heights so that the machine is nearer at hand for servicing and "keeping an eye on." Whatever the tower height, there is no doubt that the life of a wind generator left in turbulent winds will be considerably shorter than one located in a smooth windstream.

Wind Measurement

The first step in planning for a wind system is to find out the average wind speed over the site you have in mind. The most reliable way to do this is to buy or rent a wind-measuring device, an anemometer.

Where a low-cost wind system is planned, the expense of an anemometer may be avoided by using U.S. Weather Bureau records,
or applying the Beaufort scale.

The details of wind flow over any given area can be obtained
from the National Climatic Center (as noted earlier). The most that
can be gained from these records, however, is the general wind
pattern for a given locality. One is then left to deduce if the same
pattern is to be found on the specific site in question. This is done by
comparing conditions at the official recording site and at the intended windmill site. Most weather records are taken on open sites
at a height of at least 30 feet above ground and 20 feet above the top
of any obstacle within a 300-foot radius. If there is a distinct similar-
ity between the two sites' conditions, then one can be confident in
using the National weather records. A number of agencies other
than the U.S. Weather Bureau also may collect similar local data —
National Park Service areas, airports, military installations, air pol-
lution agencies and some schools and colleges.

More often than not there will be a difference in the lie of the
land between your site and the recording site, and it is difficult
without an instrument to judge the difference in terms of wind flow.
An accurate indication of wind speed at any given moment can be
gleaned from the Beaufort scale (see page 12). Two small and inexpensive wind measuring (but not recording) devices are manu-

Photo 3 (left): Dwyer hand-held wind meter.
Photo 4 (below): Dwyer windspeed indicator.

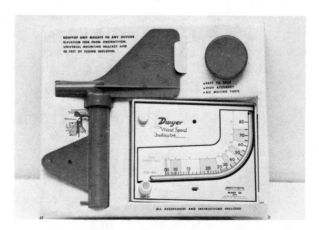

factured by Dwyer Instruments (see Manufacturer section). The first is a small hand-held wind meter. The second, a wind speed indicator, can be mounted up to 50 feet away from its indicator panel. By using the Beaufort scale or one of the two devices, a general assessment of wind speed may be made and compared with the daily records of a local weather station. Obviously if winds on a particular site are persistently strong and not too turbulent, there is an operational justification for the purchase of a small, inexpensive generator — such as the Winco Wincharger — suitable for a low-energy system.

Using An Anemometer

However, enough of this vagueness. Let us turn our attention to the use of anemometers, which are not really all that expensive — a good one costing around $100. An anemometer records on a counter the total run of the wind past its position. By reading the counter at the beginning and end of any period — whether a day, week or month — the average wind speed during the period can be accurately calculated. It is essential to use an anemometer if details of the exact wind power available are required. Also, it is advisable to use one if the site is obstructed or not open to winds from all directions. The use of an anemometer in such conditions will avoid any later disappointment once a wind energy system is installed.

Photo 5: Sencenbaugh cup anemometer.

The Beaufort Scale of Wind Speeds

Beaufort Number	Description	miles/hr.	Effect on land	Effect at sea
0	Calm	Less than 1	Still: smoke rises vertically	Surface mirror-like
1	Light air	1–3	Smoke drifts	Ripples form
2	Light breeze	4–7	Wind felt on face, leaves rustle	Small, short wavelets, not breaking
3	Gentle breeze	8–12	Leaves and small twigs move constantly, streamer extended	Large wavelets beginning to break, scattered white horses
4	Moderate breeze	13–18	Raises dust and papers, moves twigs and thin branches	Small but longer waves, fairly frequent white horses
5	Fresh breeze	19–24	Small trees in leaf begin to sway	Moderate waves, distinctly elongated, many white horses, isolated spray
6	Strong wind	25–31	Large branches move, overhead wires whistle, umbrellas hard to control	Large waves begin with extensive white foam crests breaking, Spray probable
7	Moderate gale	32–38	Whole trees move, offers some resistance to walkers	Sea heaps up, white foam blown downwind
8	Fresh gale	39–46	Breaks twigs off trees, impedes progress	Moderately high waves with crests of considerable length, spray blown from crests
9	Strong gale	47–54	Blows off roof shingles and chimney pots	High waves, rolling sea, dense streaks of foam, spray
10	Whole gale	55–63	Trees uprooted, much structural damage	Heavy rolling sea, white with great foam patches, very high waves, much spray reduces visibility
11	Storm	64–72	Widespread damage (rare inland)	Extraordinarily high waves, spray impedes visibility
12	Hurricane	73–82	—	Air full of foam and spray, sea entirely white.

Photo 6: Sencenbaugh anemometer (odometer) readout panel.

The way to assess wind patterns is to compare anemometer results
with the detailed readings taken at a local weather station. The anemometer should be read daily or weekly for about three months.
If readings can only be taken once a month it is better to record
during six months. Next, the readings for each period are compared
with those of the weather station, and a plus or minus relationship
between the two sets of figures is ascertained. Once this correction
factor is taken into consideration, then the full wind spectrum on
your particular site is known. The most important fact to be gained
from the weather station records is the *duration* of given wind

Figure 7. Variation in wind speed, shown as a fraction of the annual average.

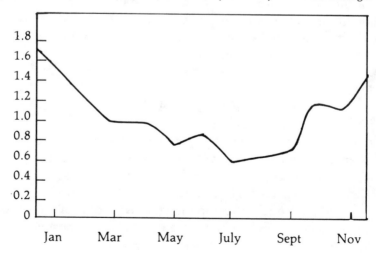

speeds, which in turn will give a clear idea of what energy one can expect from any make of wind generator placed on that site.

There is a second way of estimating wind energy in locations where comparison with a weather station is not practicable. Details are given in the next section.

Knowledge of seasonal variations in wind speeds is very useful in planning your wind energy system. Knowing the seasonal variations means knowing the energy pattern for a windmill system and planning accordingly. It is good that the wind blows hardest when we need it most, in the dark and cold of winter.

Wind Energy Estimation

As explained earlier, wind power goes up as the cube of the wind speed. That is, double the wind speed and an eightfold increase in power is gained. There is a second rule: Double the diameter of the propeller and the power obtained is increased by a factor of four. These two rules govern wind power, and they are only limited by a third, Betz's Law, which follows:

Energy is obtained from the wind by slowing down the air. The windmill cannot extract all the energy; otherwise the wind behind the propeller would come to a standstill. Betz's Law states that a windmill derives maximum power when the wind is slowed to one third of its initial velocity, in which case the power extracted is 0.593 of the maximum potential.

Based on this law, *useful* power in the wind may be expressed as:

$$P = 0.0031 \, A \, V^3$$

Where P is power in watts, A is the area swept by the propeller in square feet, and V is wind velocity (speed) in mph. This equation is based on an air density of 0.08 lb./ft.^{3}, which holds good except for slight variations in high altitudes.

To avoid the trouble of computing the swept area in square feet, the same equation can be expressed more simply for most windmills as:

$$P = 0.0024 \, D^2 \, V^3$$

Where D is the propeller diameter in feet. The two rules of diameter squared and velocity cubed can be seen in the above equation.

Using that equation, the maximum power available to a 16-foot diameter propeller in a 12 mph wind is expressed as follows:

$$
\begin{aligned}
P &= 0.0024 \times 16^2 \times 12^3 \text{ watts} \\
&= 0.0024 \times 256 \times 1728 \\
&= 1{,}062 \text{ watts (or } 1.062 \text{ kw)}
\end{aligned}
$$

However, inefficiencies within the windmill must be taken into consideration and deducted from the theoretical maximum power, shown above. A well-made propeller (or rotor) will extract 70 percent of available energy. If there is a gear between the propeller and generator, its efficiency should be at least 95 percent. The generator itself should be no less than 75 percent efficient. It is safe to assume an overall wind generator efficiency of 50 percent ($0.7 \times 0.95 \times 0.75 = 0.5$). Therefore, the actual electrical energy available may be expressed as:

$$P = 0.0024 \, D^2 \, V^3 \, 0.5 \text{ or } 0.0012 \, D^2 \, V^3$$

And so the *actual* output from a 16-foot propeller in a 12 mph wind is reduced from a theoretical 1,062 watts as above, to 531 watts.

Multiplying the average wind speed by the windmill's output at that speed will *not* give a true reflection of the average monthly or annual output. To arrive at such a figure correctly, the average windspeed must be broken down to the more detailed figures available from the National Climatic Center. For example, let us take a site some 30 miles away from a weather recording station. An anemometer is used for 6 months and the average wind speed is found to be 12 mph at a height of 30 feet. From comparison, the wind speed at your site is 5 percent higher than at the weather station, which records an average wind speed of 11.5 percent. This difference is not very important, especially as it is in your favor. However, a 5 percent correction factor is included in the calculations that follow.

TABLE 2

Energy Estimation for a 2 kw Wind Generator

No. hours per month	Weather Station Wind Speed (m.p.h.)	5% site adjustment	2 kw wind generator output in kilowatts	Monthly output in kilowatt hours
272	0–7	—	—	—
175	8–12	8–12	0.4	70
188	13–18	13–19	1.1	206.8
64	19–24	20–25	1.9	121.6
24	25–31	26–32.5	2.0	28
6	32 upwards	33 upwards	2.0	12
				438.4 kwh

While the estimating of energy available from the wind is not an exact science, the method above is reasonably accurate.

The number of hours that the wind blows at given wind speeds is supplied by the weather station. In the third column is the 5 percent correction factor for the difference between the two sites. In the fourth column is shown the output of a 2-kw wind generator at those wind speeds. The manufacturers of all wind generators supply such figures, and if they don't they certainly should. The monthly output in kilowatt hours is easily ascertained by multiplying the number of hours by the kilowatt outputs in column four.

The anemometer on this site was erected at a height of 30 feet, and it is at this height that the output of 438 kwh/month can be expected from the 2-kw wind generator. If the same generator is placed on a 60-foot tower, then we will have to revert to Figure 1 (under site selection) to find the power increase factor for the new height. At 30 feet there is about three times the power as at 5 feet; at 60 feet there is 4.4 times the power. By going from 30 to 60 feet on this unobstructed site, a power gain of 4.4 ÷ 3 or 1.47 is achieved. Hence, the expected output of 438 kwh is increased to 644 kwh simply by adding 30 feet of tower.

Second Energy Estimating Method

There is a quick and simple method, based on tests of 23 small wind generators, that gives an *approximate* indication of the power potential on sites for which only the average wind speed is known. The curves drawn on Figure 8 are not definitive but reflect a general pattern. The cut-in speed on the 20 mph line is 10 mph, on the 25 mph line, 13 mph and on the 30 mph line it is 17 mph. The shut-down speed for all three lines is taken to be 60 mph.

Let us return to the 12 mph average wind speed example used with the first energy estimation example. The 2 kw wind generator used in the example had a cut-in speed of 8 mph and a rated wind speed of 25 mph. Using Figure 8 again, an annual output of 1,700 kwh per kilowatt can be expected from an average wind speed of 12 mph. Output from the 2 kw generator would, therefore, be 3,400 kwh per annum or 283 kwh per month. To this must be added 70 kwh, as the cut-in speed on Figure 8 is 13 mph, whereas the 2 kw generator cut-in speed is 8 mph. So, using the graph, we arrive at a total of

Figure 8. Annual output in rwh per generator kilowatt for various average windspeeds. Outputs for three wind generators with different rated windspeeds are shown.

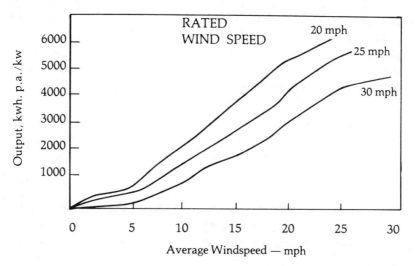

353 kwh, which does not compare too badly with the original of 438 computed by Table 2.

The whole question of assessing windmill output from an average wind speed is now being made easier by manufacturers who are beginning to include such estimates in their catalogs, though there is a reluctance on the part of some to do this. They believe, and rightly so, that the characteristics of individual sites have an important affect on the wind generator output. But this need not stop them from reporting the actual output of a windmill and describing the site characteristics so that others may make allowances for differences.

It should be mentioned here that net output from the wind generator is not the final power available for use. Unfortunately, there are transmission line, battery and inverter losses to be deducted. These, and other wind energy systems we will discuss in the next section, but first a look at the various types of windmills which can be used to turn a generator.

Windmills

Strictly speaking, a "windmill" is used for milling (grain grinding, etc.), and not for the generation of electricity or water pumping. But, being short and evocative, the word tends to be used to describe what is more properly called any of the following: wind generator, wind-driven turbine, windplant and wind machine, to mention but a few.

There are basically two types of windmills:

Horizontal axis. Where the propeller (or rotor) on a horizontal shaft or axis moves in a plane perpendicular to the direction of the wind. This includes the multi-blade, four-arm and high-speed propeller types;

Vertical axis. Where the rotor on a vertical shaft or axis has its effective wind-catching surface moving in the direction of the wind. This includes the more recently developed Darrieus and Savonius rotors.

Horizontal Axis Windmills

Propeller-driven, horizontal-axis windmills come in all shapes and sizes, but there is one characteristic which differentiates the various types. That is their tip speed ratio, which is the ratio between the propeller tip speed and the wind speed. This ratio can range from 1:1 for slow-speed mills, up to 8:1 for modern high-speed propellers. Mills with low ratios are mainly suited for slow-speed purposes such as water pumping and other mechanical uses. High tip speed propellers are suitable for generating electricity.

$$\text{Tip speed ratio} = \frac{\text{Speed of rotation of blade tip}}{\text{wind speed}}$$

$$= \frac{2 \, \pi \, R \, rpm}{88V}$$

Where R is radius, *rpm* is the revolutions per minute of the propeller and V is the wind speed (velocity)

Example. A six-foot diameter propeller rotates at 700 rpm in a 23 mph wind, the tip speed ratio is $2 \times 3.14 \times 3 \times 700/88 \times 23$ and equals 6.5.

A tip speed ratio of 6.5:1 means that in a 23 mph wind, the propeller tip travels at 150 mph. If the ratio were 1:1, the propeller tip speed would be 23 mph.

Slow-speed, high-torque mills generally have propellers with high "solidity" ratios. That is, the area they sweep is "solid" with blades or sails. They usually have a low power coefficient, as the multitude of blades create air turbulence and cause a negative drag. Nevertheless, they are safe, suit the mechanical purpose for which they were designed, and, as a result of their high solidity ratio, they produce power in low wind speeds. Modern airfoil propellers have a very low solidity ratio. They produce little or no power in wind speeds under 8 mph, but they make excellent use of high wind speeds where the really useful power is. Their high tip speed ratio makes them ideal for driving generators. Indeed, the problem faced by the manufacturers of very large (100-foot-plus) propellers is to design propellers capable of holding together when spinning at the speed of sound.

The Sail Windmill

The sail windmill has its home in the Mediterranean countries — Crete in particular — where they number tens of thousands and are used for water pumping. The sail mill is probably the simplest and safest mill for home construction. Despite its low tip speed ratio of 0.75:1 it has many advantages. Being made from wood and cloth, it is inexpensive and easy to repair. Its slow speed and high solidity make it responsive to low wind speeds, and with a maximum speed of about 50 rpm, it is safe. Moreover, at that sort of speed the white revolving sails set against a blue sky are a delight to watch.

The sail mill is, as its name implies, made from sailcloth that is attached to a number (usually eight) of wooden spars. The solidity ratio varies from 0.1 as in photo 9, to 0.6. The lower ratio applies in high winds, when the sails should be reefed (wrapped around the spars) to avoid tearing. The whole rotor structure is strengthened by tie-wires extending from a forward extension of the shaft, called a "bowsprit." Besides its suitability for mechanical purposes, the sail mill can be geared up to generate a small but useful electrical output. A speed-increasing gear of between 30 and 40 to 1 is needed if it is to drive an automobile generator. That's a cumbersome ratio, but it can be done using gears and one- or two-stage belt drive.

Home construction of a sail mill can be very inexpensive, especially if second-hand and ideally suited auto parts — such as shaft, brake drum, gears and alternator — are used. The sailcloth and wooden spars are also inexpensive. But remember that where the cost in cash is low, the cost in time tends to be high.

The sail mill is self-governing to a degree in that it will spill excess winds by flapping, and thus lose its natural airfoil shape. The big disadvantage is that this self-governing action is of little use when the wind is excessively high. In such winds it is necessary that the rotor's solidity be reduced, which in turn greatly reduces the force acting on the rotor and its support. Reducing the solidity means reefing the sails — pleasant work in summer, but a totally different matter if you have to get up in the dark of a winter's night, stumble through the pouring rain to reach the mill and struggle to reef the sails.

Actually, this sounds worse than it is. With a good knowledge of local wind speed, this crisis need rarely happen, and if it does the damage done is usually just a few torn sails. Provided the demand for power is low, the sails may be left partially reefed throughout the winter season. As an alternative to reefing, a simple hook arrangement can be made for fixing and removing separate sails, for example leaving two instead of eight sails operating in high winds.

While there is little written on the sail mill, compared to the high speed types, there are available four sets of reasonably good home construction plans, detailed below (see *Bibliography* for their publishers):

Do-It-Yourself Sail Windmill Plan describes how to build the 12-foot diameter sail-driven generator shown in photograph 9. Starts charging an 8 mph wind and generates a useful 200 watts at 15 mph.

Photo 9: Sail wind generator, a "soft" technology that generates up to 300 watts.

Photo 10: Forty-foot sailcloth windmill on a brick tower.

Reinforced Brickwork Windmill Tower. Despite its title, it gives plans for construction of the 40-foot diameter mill and the brick tower shown in photograph 10. Is used for pumping water, starts operating in winds of 3-4 mph and produces a maximum of 13 hp in 25-30 mph winds.

25 Foot Diameter Sail Windmill. A design manual prepared by Windworks for the Brace Research Institute. Produces power at 5 mph and a maximum of 6.7 hp at 20 mph. Has six instead of eight sails.

Food From Windmills. Describes in detail the construction and operation of 11-foot diameter sail water pumping mills built by the American Presbyterian Mission in Etheopia.

The overall efficiency of a sail-driven generator is at least 15 percent and can be as high as 25 or 30 percent, which is not bad considering the average 50 percent efficiency of high-speed propeller-

driven generators. I am not too concerned about theoretical "efficiency" when it comes to wind generators. Surely a wind machine
is efficient if it suits the purpose for which it was designed and does
not waste resources. It seems futile to expend great effort in designing a super-efficient propeller when the same result can be obtained simply by increasing the propeller diameter.

Multi-Blade Mills (or Fans)

There are few who have not seen one of these old water pumping
mills, for they still dot the countryside, some in action but many
in ruin. The 19th Century saw hundreds of thousands of such mills
in operation on farms across the country. Indeed there are companies which have continued to manufacture them from those
days right up to the present (photo 11).

*Photo 11: A multi-blade windmill used now to pump water through an array of
solar collectors.*

Photo 12: A 25-inch diameter venti-lation fan with seven instead of four-teen blades that will produce 50 watts in a 35-mph wind. With 14 blades in place, both speed and output will be lower. This fan is intended for marine use.

The swept area of many multi-blade mills is more or less "solid" with blades. In other words they have a high propeller solidity. Therefore, they are slow speed, having a tip speed ratio of less than 1:1 — similar to the sail mill, though regarded as more efficient. The multi-blade converts low wind speeds into useful torque power suitable for water pumping and other mechanical purposes.

Four-Arm Mills

This is the classic windmill of which few remain, and those which do are generally "retired." Some of these old mills were remarkably advanced in propeller design, efficiency and strength. At the beginning of the last Century, propellers 100 feet in diameter were frequently used in Holland. Some of the old Dutch mills have recently been converted to generate electricity, one of them originally built in 1727.

The Dutch four-arm has made a comeback at Santa Nella in California. There, a 46-foot diameter mill rotates on top of the "Pea Soup Anderson's Restaurant." In a 20 mph wind, frequent in that area, the mill generates a useful 8 kw of electric power. Designed by Wind Power Systems, Inc. (see *Manufacturers*), it is capable of withstanding gusts of up to 120 mph with the blades stopped. Automatic brakes hold the propeller in winds over 25 mph.

Photo 13: The "retired" four-arm still retains its grace.

Photo 14: This recently built "Dutch" four-arm generates a useful 80 kw at an overall efficiency of 20 percent.

Advanced Propeller Design

Most commercially available wind generators today use two or three long and slender blades. The design of the blades is such that they produce maximum lift and minimum drag. It is as result of this "lift" that the blades attain their exceptionally high tip speed ratio, between five and ten times the wind speed. This, together with an efficiency as high as 70 percent, makes the two- or three-bladed propeller well-suited to the generation of electricity (photo 15).

The very low solidity ratio of two-bladed propellers gives them a slightly higher tip speed ratio and aerodynamic efficiency than their three-bladed counterparts. Twin blades, however, are rarely used on wind generators with outputs in excess of 1 kw. The reason for this is vibration, which so frequently sets in. I know of one person who fixed a two-bladed Winco 200 watt to his stone house, and even though the walls were thick, the vibration was transmitted from the tower through the walls to shelves, where the crockery and glassware kept inching their way to the edge.

Photo 15: With a maximum tip speed of 200 mph, this little Winco 200 watt moves fast and furiously. The blades do actually bow backwards in high winds. Air flaps extend to prevent overspeed.

There seem to be two causes: First, air turbulance may be increased by a bounce or shadow effect on one blade as it passes the tower; and second, if the main tower bearing is not rigid, there tends to be a rocking of the whole wind generator.

Exceptions to the use of the twin-bladed propellers just for small generators, are the huge Smith Putnam and ERDA wind machines with propellers of 175 and 125 feet in diameter respectively. The massive Smith Putman wind generator in Vermont, which had a maximum output of 1,250 kw, lasted for several years, until one of its twin blades snapped at the root — and that was the end of that. The ERDA twin-bladed 100 kw mill has not had a very impressive record: it has been operational for only thirty-six hours. Yet there are plans to build many more such turbines. Some twin-bladed mills use two side weights to give the stability of four blades.

Photo 15A: ERDA twin-blade 100 kw mill.

For all practical purposes, the three-bladed propeller is just as good as the twin bladed, and has the added advantages of greater stability and a lower starting speed, due to increased blade solidity. Four-bladed propellers, popular back in the Thirties, have now been replaced by the three-bladed version. No matter what the blade configuration, correct blade balance is essential for stability; each blade must be exactly balanced so that the weight is evenly distributed around the hub (photo 16).

Blades are made from all sorts of materials: metal, wood, fiberglass, cloth and certain reinforced plastics. Metal blades have been known to cause some radio and television interference, but other than that they tend to be reliable. Wooden blades, by far the most popular, require particular care in their shaping if they are to give long service. The leading edge of each blade requires reinforcing with some harder material. Copper strip is often used for this purpose and is stapled into the wood, a practice I am not too fond of, since the staples can weaken the grain of the wood. I know of at least one case where rot developed in the blade as a result of water seeping in alongside the staples. All blades need a protective coating of varnish or some hard-wearing paint.

Photo 16: Three-bladed upwind propeller driving a 5 kw generator. The three spikes at the hub are speed governors. The tail vane faces the propeller into the wind.

Photo 17: An 18-foot diameter sailwing used for pumping water at the New Alchemy Institute in Massachusetts. With a gear up of 25:1, it can be used to generate electricity.

Cloth can be used as a blade or "sailwing" when it is stretched as a sock over a metal and wire frame to form an airfoil section (photo 17). This type of propeller is suitable for home construction. Plans for one are published in the *New Alchemists Journal*, Number 2 (see *Bibliography*). Much that is said in favor of the sailcloth mill can be said for the sailwing; but the sailwing has two great advantages: It has a much higher tip speed ratio and does not need reefing. In the event of extreme wind speeds, any loss which may result from flying sail material would be considerably less than that sustained by wood or metal.

Vertical Axis Rotor Windmills

The chief advantage of the vertical axis rotor is that it does not need orientating into the wind. It is omnidirectional. A vertical axis Savonius rotor may be made by cutting an oil barrel in half and placing the two halves as in Figure 18.

The Savonius rotor has a slow speed, and with an efficiency of

Figure 18: A home-built, vertical-axis, oil barrel rotor.

about 20 percent is mainly suited to water pumping, though there are a number of people who use the rotor to generate a low electrical output.

The Darrieus vertical axis rotor and its many variations have a far greater potential for the generation of electricity than the Savonius. In fact, there are three different Darrieus-type wind generators available today.

Graph 19. Tip-speed ratio vs. power coefficient for various windmills. Each type of windmill operates most efficiently at a particular windspeed.

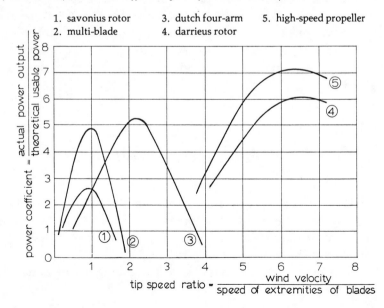

Photo 20: Darrieus vertical-axis rotor.

The Darrieus has speed and efficiency characteristics similar to that of high speed propellers (see Graph 19). All Darrieus rotors have two or more long, thin blades, which may be curved (like a hoop) or straight, that rotate around a vertical shaft. Originally developed back in the Thirties, many such vertical rotors still have a space age look about them.

Figure 21. Vertical-axis wind turbine. Front view, above, and plan view, below. (From Clegg, Low-Cost Sources of Energy for the Home, *1975.)*

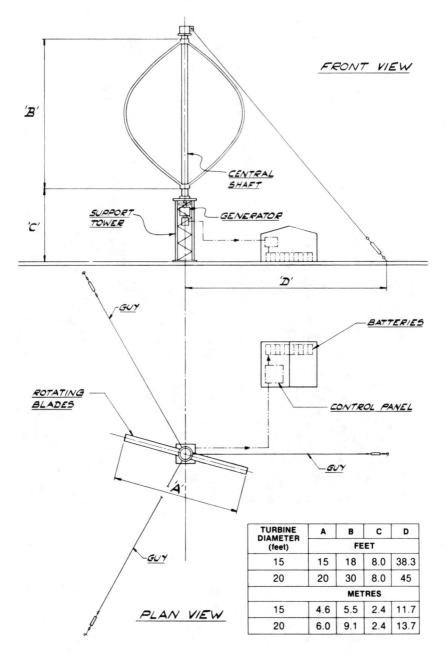

FRONT VIEW

'B'

'C'

CENTRAL SHAFT

SUPPORT TOWER

GENERATOR

'D'

GUY

BATTERIES

ROTATING BLADES

CONTROL PANEL

GUY

'A'

GUY

PLAN VIEW

TURBINE DIAMETER (feet)	A	B	C	D
	FEET			
15	15	18	8.0	38.3
20	20	30	8.0	45
	METRES			
15	4.6	5.5	2.4	11.7
20	6.0	9.1	2.4	13.7

The DAF rotor manufactured in Canada, produces 4 kw at 25 mph with a 15 ft. rotor, or 6 kw with a 20 ft. rotor (see Figure 21). Darrieus rotors are not self starting. They require a "push" from a motor to get going. With the DAF this happens every 15 minutes. If the wind is strong enough the rotor will continue spinning (a most unusual sight), and if not, it will stop. This starting system seems wasteful and could be much improved by the introduction of a wind-pressure switch, which would start the rotor when the wind speed warranted it. Another disadvantage is that the guy wires that extend from the top of the shaft to the ground very much restrict the tower height. But then again, this becomes a theoretical consideration of no importance if the rotor operates successfully and at a reasonable price. The standard 8-foot tower makes the generator accessible from the ground.

The Cycloturbine, developed by Pinson Energy Corporation, is a straight-bladed Darrieus which, as a result of its variable pitch blades, has the singular advantage of being self-starting. The tail or vane at the top of the rotor shaft (see photograph 22), senses the wind direction and adjusts the blades accordingly.

Photo 22: The variable-pitch Cyclo-turbine, rated at 4 kw at 25 mph.

Another interesting and recent development is the twin-bladed variable pitch rotor shown in Figure 23. The twin blades are hinged to the cross arm and allowed to vary their pitch to suit changing wind speeds. This also avoids the extreme bending stresses caused by high rotational speeds. It is because of these stresses that the DAF rotor has a curve (called a troposkien), which is the same shape as that obtained by rotating a rope about a vertical axis. The expense of fabricating the curve is avoided in this double-bladed rotor by using straight blades. Unlike the Cycloturbine, however, the rotor is not self-starting. Neither of the variable-pitch rotors requires restricting guy wires as the DAF does.

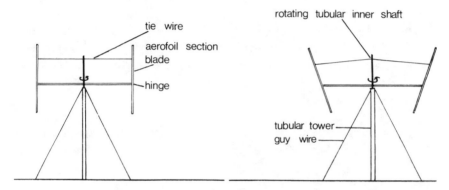

Figure 23. Variable-pitch rotor configurations at low windspeed (left) and at higher windspeeds (right).

An arrangement similar to the propeller sailwing may be used on a vertical-axis rotor (see Figure 24). The sailwing, held in tension, achieves a variable-pitch airfoil profile as it rotates.

Another recent development in wind power is the rotor shown in photograph 25. Each blade is twisted in its length through 180 degrees to form a helix, and the rotor is self-starting. Although the tip speed is similar to the wind speeds, the rotors have the high aspect ratio of four (length to diameter), and therefore have a relatively high shaft speed. The model shown consists of two helical rotors, each 8 by 2 feet and capable of generating 1 kw at 28 mph. Figure 26 shows a similar rotor, which uses sailcloth, developed by Zephyr and called the "Tetrahelix."

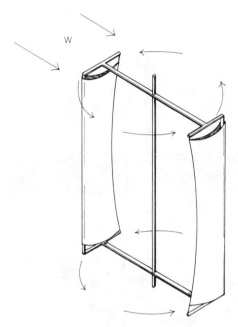

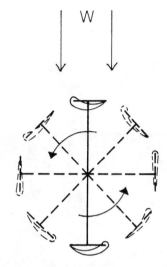

Figure 24. Vertical-axis sailwing, showing view from above.

Photo 25: An alternative twist to wind power—the helical double rotor.

Figure 26. The Tetrahelix.

Photo 27: A rotor-driven boat.

Photograph 27 pictures Mr. A. Flettner standing in his rotor-driven boat. It is said that he succeeded in crossing the Atlantic in a rotor-driven ship in 1925! He also suggested that the rotors (which are cylinders spinning about their axis) could be used as blades for windmills.

Flettner rotors operate on the "Magnus effect" — when a cylinder spins at a sufficiently high speed, the windflow around the cylinder is unsymmetrical. A pressure then is exerted on the cylinder in a direction perpendicular to the wind, so the cylinder acts like an airfoil.

Governors

Governors must be used to prevent propellers or rotors from over-speeding, for a high-speed wind machine without an efficient governor is a danger to life and limb. This is one of the main reasons why I do not encourage the home construction of such machines — unless one is going to put a considerable amount of time and effort into the work, and if that is the case why not go into limited manufacture, serving your own locality? The same initial effort is required whether one or twenty-one machines are built. The more local manufacturers the better, but please do not underestimate the considerable effort, time and study required to build a reliable windmill.

To illustrate the problem of governing, let us take an example of a twin-bladed propeller, with a tip speed ratio of 7:1, which is geared up to a generator. Assume that this mill has been built from one of two sets of plans (one still currently available), neither of which bothers to mention governors. At a wind speed of 25 mph all is well, power is generated and the propeller tip speed is only 175 mph. Should the wind suddenly increase to 60 mph, however, before there is time to operate manually the flimsy hand brake, the propeller tip speed will increase to over 400 mph. At such speeds any weakness or unbalance in any part of the machine will come under extreme pressure. Even the slightest unbalance in the propeller will set up invariably destructive vibration. That the gears or generator (or both) will probably disintegrate is of little importance compared to the potential hazard of blade shatter.

Fortunately all windmill manufacturers and careful home-builders are acutely aware of this problem and they build reliable speed governors into their designs. Of the many different types of governor employed today, three are most popular. The first system uses weights attached or somehow connected to the blades. These

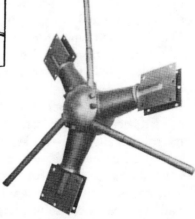

Figure 28. The Winco airbrake in normal (left) position and governing (center). The centrifugal governor (right) alters the pitch of the blades.

weights, operating under centrifugal force, alter the pitch of the blades so that they lose much of their aerodynamic efficiency in excessive wind speeds, and so maintain a constant maximum speed.

A second method is to place the propeller slightly off top-dead-center of the tower shaft. The tail vane is also hinged so that in high winds the propeller is turned out of the wind by wind pressure as the tail folds into a plane parallel to the propeller. As the wind speed declines, the tail unfolds under the force of gravity.

The third governing system utilizes air flaps or spoilers, which have a drag effect and thus act as a braking mechanism. All such flaps, whether they are fixed on the hub or the blades themselves, operate under centrifugal force.

Wind machines also have, or should have, manually operated brakes to facilitate servicing. Most manufacturers offer, as an extra, an automatic propeller brake operated by a wind pressure switch which comes into operation in winds of over about 60 mph. Since wind speeds in excess of 60 mph often take rapid changes in direction, it is very wise to have the machine shut down in such conditions. Thus an automatic brake will extend the life of a wind generator, particularly one sited in a high wind speed area.

Generators

All the old windmills back in the Thirties used slow-speed, copper-laden DC generators, and many of them are still operative today. They had the drawback that to collect the current they needed carbon brushes, which in some cases would last up to 20 years, but in others might only last one year.

To build such generators now would be prohibitively expensive, mainly due to the mass of copper required. Modern AC generators (properly called *alternators*) are lighter, smaller and therefore less expensive (photo 28A). Moreover, alternators do not require brushes, even though they can be "rectified" to produce a DC current.

The difficulty with most alternators is that they require a very high rotational speed — at least 18,000 rpm (US)/15,000 rpm (UK). To take the propeller drive of 300-400 rpm maximum from the shaft to the alternator, some form of speed-increasing gear is required. Gears cost money, add weight to a wind generator and waste what would be otherwise useful energy. Moreover, if they are not over-designed they may become the weak link in a wind system. In reaction to this a number of manufacturers have developed special slow-speed alternators suitable for direct drive from the propeller.

Photo 28A: Rewound industrial alternator, 1.5 kw at 1800 rpm, developed by Natural Power.

Towers

Earlier, under "Site Selection" the benefits of increasing the tower height were pointed out, and here we look at the various types of towers available. An example of the cheapest type of tower or mast is shown in photograph 29, where a telephone pole is used to support an old and heavy Jacobs mill. The 70-ft. pole is tied with guy wires and hinged at the bottom to a concrete base. As a result of the hinge the whole tower, with the Jacobs on top, may be lowered and raised by means of a winch for servicing. It is also good to know that should a hurricane threaten, the mill can "lie low" in safety. The main difficulty with wooden poles is that the portion left underground eventually rots, but with this system that need never happen. Street lamp posts may be used as towers, too, but it is of the utmost importance to understand the pressures acting upon the tower and to ensure that it is capable of taking such pressure.

Photo 29: Jacobs windmill supported by telephone pole.

Photo 30: Self-supporting tower with 3.5 kw direct-drive Elektro.

Most people prefer to purchase specially built towers from the wind generator manufacturer who understands what is required in the way of propeller clearance, etc. Self-supporting or guyed towers may be used. The former tend to be more expensive, but obviate the need for wires (photo 30). The tower base should be well encased in a concrete cube according to the detailed instructions supplied by manufacturers, and which differ with each type of tower.

Towers are assembled on the ground and then erected by means of a mobile crane or a gin pole. You will need friends around to assist with the gin pole method (Figure 31), and the more the merrier. Make a party of it. The tower base is set against a backstop, behind which is the gin pole. With bulkier self-supporting towers the gin pole may be set further back from the backstop. The gin pole should be a good length, at least 35 percent of the tower height. People are required to give initial lift, but after that a car or more people should finish the job. Once erected, the tower is secured according to the manufacturer's specifications.

The next job is to fix the wind generator on top of the tower. This work is made easy with a crane, but aside from the expense, cranes cannot always go where wind generators are sited. In many cases, where the machine is not too heavy, it is possible to fix the windmill on the tower top before it is raised (photo 32). This is always the case where a hinged tower is used. In some other cases the manufacturers supply a small crane which is fixed to the top of the tower and used to haul the generator to the top, where it is swung into place. *Wind Power Digest* Number 5 (see *Bibliography*) gives details on how this is done.

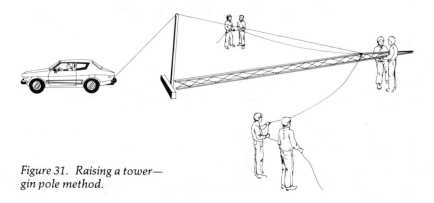

*Figure 31. Raising a tower—
gin pole method.*

Photo 32: The slender, guyed tower, complete with 5 kw Elektro, is raised from the ground by a crane.

The tower should be accurately vertical. Otherwise output may be impaired due to a tendency for the wind generator to list a little out of the wind. All towers need lightning conductors, and if there are guy wires it is advisable to ground those as well. Electric cables carrying current from the generator to the house also should be grounded before they enter the house.

Home Wind Equipment

Electrical Terms

We are going to be talking a lot about volts, amperes and watts, so let's get a thorough understanding of what these terms mean.

Voltage. Voltage is a measure of the pressure under which electricity flows. The lower the pressure the slower the flow, and any voltage (V) under 50 V is regarded as low. Car batteries operate under a pressure of 12 V, truck batteries sometimes at 24 V, and both can give a mild shock. Higher voltages, such as the common line load of 120 V, can give a nasty shock, particularly if the skin surface is wet, and contact with 240 V lines, used for heavy appliances, is to be avoided.

No home wind energy system will ever need to use voltages higher than 120 V (240 V in the UK), which is just as well as contact with such high voltages often causes fatal shocks — what a way to begin a section on home energy! In truth, there is no danger of such an occurrence provided there is no human contact with any live wire — certainly not with anything over 50 volts.

Amperage. Whereas voltage is the pressure, amperes (or amps) indicate the flow rate of that electricity. Amperes can range from a fraction of an amp to the several hundred amps required to start a car's engine. The greater the flow of amps through a wire the wider its diameter should be — much the same as water flowing through a pipe. The cause of most electrical house fires is that an excess of current (amperes) is drawn through a wire too small to carry the load. The excess is frequently the result of multi-socket adaptors, for example two 25 amp heaters plugged into one 25 amp socket. The result is that the wire overheats and may cause fire.

It is to prevent this type of occurrence that fuses are used, for when an excess of current is drawn through a fuse, it will simply melt. Replacing a blown fuse with an over-rated one defeats the purpose and causes house fires, too. The golden rule: *Never draw more current through a wire or cable than it is rated for.*

Wattage. Watts are a measure of the amount of power used by an electrical appliance. Multiply the amps at which the appliance operates by the voltage and you have its wattage. So:

$$\text{Watts} = \text{Amps} \times \text{volts}$$

from which follows:

$$\text{Amps} = \frac{\text{watts}}{\text{volts}}$$

and

$$\text{Volts} = \frac{\text{Watts}}{\text{amps}}$$

Hence a 12 V car light bulb which draws two amps is a 24 watt bulb.

But take another 24 watt bulb, operating at a voltage of 120 and it will only draw 0.2 amps. At 0.2 amps a much lighter wire may be used to carry the same 24 watts of power under the high pressure of 120 volts as opposed to 12 volts. Running four 100-watt light bulbs at 120 volts would require only 3.3 amps, whereas the same wattage at 12 volts would draw a current of 33 amps, requiring a heavy wire.

If a 100-watt bulb is left on for one hour it will use 100 watt-hours, and if it is left on for 10 hours it will use one kilowatt-hour (1,000 watt-hours equals one kilowatt-hour). Batteries are usually rated in amp-hours. Hence a 100 amp-hour battery will give one amp for 100 hours, or 100 amps for one hour. The amp-hour rating by itself is not very informative unless the battery voltage is known. In other words a 100 amp-hour battery at *6 volts* will give only 600 watt-hours, whereas a bank of 100 amp-hour batteries rated at *120 volts* will give 12,000 watt-hours or 12 kilowatt-hours (kwh). A kwh is the standard electrical unit of measure.

Resistance. There is one other aspect of electricity which wind power workers must understand, and that is Ohm's Law. The ohm is a measure of any material's resistance to the flow of electricity. Materials with very high resistance, such as plastic, are used as insulation. On the other hand metals such as copper or aluminum have very low resistance to the flow of electricity and are used as electric conductors.

At low voltages, such as 12 or 36, a loss of power can easily occur where a wind generator is sited far from the house. Thick copper cable suitable for conducting 12 volt current is expensive, and very expensive if one has to buy more than 100 or 200 feet of it. But what is possibly worse than the cost is the loss of power due to resistance in the thick wire.

The resistance loss of copper and aluminum wire is shown in the accompanying table.

All electric circuits have two wires, one positive and the other negative. The voltage drop in a wire is equal to amps times the resistance of the wire: Voltage drop = Amps × Resistance.

Wire Gauge (A.W.G.)*	Resistance in ohms per 100 ft. (two wire)	
	Copper	Aluminum
000	0.0124	0.0202
00	0.0156	0.0256
0	0.0196	0.0322
2	0.0312	0.0512
4	0.0498	0.0816
6	0.079	0.1296
8	0.1256	0.206
10	0.1998	0.328
12	0.3176	0.522

* American Wire Gauge.

A drop in the voltage (due to resistance) will have a damaging effect on most electrical equipment. The power loss in watts in a wire is equal to amps squared times resistance: Power loss (watts) = Amps² × Resistance.

And if that is as clear as mud to you, I am not surprised. It was the same to me for a long time. However, the following example may help clarify the matter:

Power produced by a wind generator is 600 watts, and 200 feet of No. 4 (AWG) copper wire is used to carry power to the battery bank. Wire resistance (see above), equals 0.1 ohm (0.0498 × 2 = 0.1). The choice of generator voltages is 12, 24, 32, 120 or 240.

Power Generated	Generator Voltage	Generator Amps Output	Line Voltage Drop	Voltage at load	Power Loss in wire	Power at load
600 Watts	12 Volts	50 Amps	5 Volts	7 Volts	250 Watts	350 Watts
600	24	25	2.5	21.5	62.5	537.5
600	32	18.75	1.87	30.12	35.15	564.85
600	120	5	0.5	119.5	2.5	597.5
600	240	2.5	0.25	239.75	0.62	599.37

In this particular case it can be seen that doubling the generator voltage cuts to a quarter the power loss. There will always be some power loss in an electric line, but the object of understanding Ohm's Law is that the voltage and wire gauge may be chosen so as to minimize losses.

Wind Systems

The two different *types* of electricity are *direct current* (DC) and *alternating current* (AC). AC current is supplied by the power company at a voltage of 120 V (240 V in the U.K.) to provide the type of current and voltage which all domestic appliances operate on. Moreover current from the power company is supplied at a frequency of 60 Hertz (50 Hertz in the U.K.). Any variation in this frequency causes appliances such as TV's and stereos to "brown" or "black" out, but resistance appliances such as heaters will operate as normal on varying frequency (and voltage).

But because of ever-changing wind patterns it is impossible for wind generators to produce a constant voltage and constant frequency electricity for use in the home. Therefore AC current is taken from the generator for heating, or as DC for battery charging.

In many cases the AC current from a wind-driven alternator is "rectified" to become DC, since AC current cannot be stored but DC can. The DC battery is an important link in a wind energy system. With a battery, excess power generated in high winds may be stored for use at any time. DC current from the battery can be used in DC house circuits similar to those used in cars, or it can be changed to AC current, at any pre-selected voltage or frequency, by means of an inverter.

Battery Storage

Stationary type batteries are generally used with wind systems, although car batteries, new or second-hand, certainly can be used. But they may be more trouble than they are worth, for they are designed to give a high charge for a short period, whereas stationary batteries give a steady output over a prolonged period. Deep-cycle, lead acid stationary batteries have extra-heavy plates and special separators. Life expectancy is 10 to 20 years with an efficiency of between 65 to 85 percent.

Batteries become less efficient when they are cold, so most people choose to keep their battery banks indoors. A word of warning on this: Batteries give off hydrogen gas, and if this gas is allowed to build up in a closed, unventilated area, it may explode. The explosive nature of hydrogen has hindered the development of hydrogen as a cheaper means of energy storage than batteries. For it is possible to use a wind generator and by the process of electrolysis transform water into hydrogen and oxygen.

Stationary batteries are generally used in other applications for emergency lighting systems, golf-carts, industrial trucks, etc. Nickel cadmium batteries may also be used in a wind power system. New they are very expensive, but second-hand they may be a good buy. Efficiency is only 50-70 percent, but they are suitable for total and high rates of discharge, and their life expectancy is in excess of 20 years (photo 33 below).

Battery Bank Size. The needed size of the battery bank depends upon two factors: the local wind pattern and the allowable charg-

Photo 33: A battery bank capable of storing up to 12 kwh.

ing rate. The wind pattern establishes how long and frequent are windless periods over the year. In general four to seven days' energy needs should be met by the battery bank, and more than that would be a costly extravagance. It is possible that a period of 10 windless days will occur anywhere, but it is more cost-effective to cover this peak storage need with a stand-by generator.

An important characteristic of batteries is that they should not be charged too rapidly, or their life expectancy will suffer. Maximum charging rate is 14 amps per 100 ampere-hours storage, and this gives a rough way to assess the size of the battery stock required. In practical terms this means that even where very favorable wind patterns indicate a minimal battery stock is needed, it nevertheless has to be big enough to take into account the charging capacity of the windmill. A voltage regulator is also required to stop the wind generator from overcharging the batteries, except in cases where the charging rate is less than 5 amps per 100 ampere-hours storage.

Standard Inverters. Inverters which change DC current to AC should be used as little as possible, for they are both expensive and inefficient. There are two types: the rotary and the static.

The rotary inverter is a DC motor driving an AC alternator. The efficiency is about 50 percent to 75 percent, but it also draws a "no-load" current of 15 percent - 20 percent. A 1.5 kw rotary inverter therefore will waste up to 7.2 kwh a day, unless it is switched off when not in use. Rotary inverters are less expensive than static inverters, and when closely matched to the load there is not a great difference between the two in inefficiency. Static inverters are around 85 percent efficient and only draw a no-load current of 2 to 4 percent.

Another important aspect of inverters is the wave shape they produce, which can be square or sine wave. Cheaper inverters produce a square wave, and the importance of this is that most household appliances with motors are designed for operation on a sine wave, as is supplied by the power company.

Finally, be sure that any inverter chosen is capable of coping with the surge of power required to start motors.

To avoid the cost of an inverter, use a DC system wherever possible. DC motors are readily available, as are lights, in various voltages. The car trade uses 12 V equipment, while many boats and trains still operate on 32 volts. To find needed DC equipment can be a bit of a search, but it is worth it in the end. As a bonus, 12 V bulbs last for years and years.

Back-up Power. So by use of batteries and an inverter, normal power company-type electricity can be had from a wind generator. A small back-up generator may be used to charge the batteries in times of low wind. The generator may be driven by gasoline, LP gas or propane. (North Wind Power Co. sells conversion kits for LP gas and Propane.) The output of the back-up generator can be a fraction of the rated capacity of the wind generator, which will provide the essential power, or it can be equal to the full capacity.

Whatever its capacity, a stand-by generator will be called upon to supply less than three per cent of the total annual energy budget of the typical domestic wind system. The stand-by generator can be started manually or automatically to charge the batteries when they are low. A wind energy system complete with such a back-up generator provides what is probably a more reliable service than the power company. On the other hand those who can get along without electricity for a few days occasionally will get along fine without the back-up generator.

Gemini Inverters. A particularly interesting device called the *Gemini Inverter,* is now available to use wind generators in association with a public utility power line. The Gemini links the output from the wind generator to the power line and the domestic circuit. The great advantage of this method is that the varying voltage and frequency generated by the windmill is instantly converted to exactly the same type of electricity distributed by the utility's power grid. Thus your wind power is fed to the company, which feeds back power to you when you need it.

This is accomplished without the aid of batteries or standard inverter, but does depend upon having a local power line connection and the permission of the power company. Many power companies are helpful in these arrangements. The only drawback is that they tend to give less credit for excess power fed into their line than they charge for additional power you use. Should complete freedom from

the power company lines be desired, a battery and standard inverter system may be used.

For more on Gemini Inverters, see page 61 and 110-111.

See also PDI's "Base Load Injector," page 102.

Examples and Costs

Where a house is not adjacent to power company lines, the choice
of wind power is most attractive, especially since overground power
lines now may cost in excess of $10,000 per mile. A good wind
energy system will cost less.

The cost of a windmill system is a little more than that of a diesel
generator. But with the wind there are no fuel costs and no fumes,
and the sound of a windmill is far more pleasant than the noise of
a diesel engine. Where there is a reasonable wind, a wind generator
is far more economical per kwh produced than a diesel generator.
In such cases the economic argument for wind power is based on
a simple capital cost comparison basis. And if that does not come
out in favor of wind power, then a look at the cost of power company electricity and diesel oil will invariably settle the matter.

Cost Increases of Fuels*

Year	Coal	Fuel Oil	Natural Gas	Electricity
1971	1.000	1.000	1.000	1.000
1972	1.109	1.056	1.000	1.000
1973	1.242	1.578	1.000	1.033
1974	1.529	2.345	1.075	1.061
1975	2.184	2.907	1.466	1.370
1976	2.322	3.078	1.607	2.031
Avg. annual increase	26%	41.4%	12%	20.6%
Overall average annual increase		25%		

* Based on average figures for Western countries. U.S. cost increases slightly less.

There is no escape from the above figures, except by using inflation-free alternative sources of energy. Don't be lulled by the apparent low price of natural gas, which is already in short supply. When the world oil shortage arrives in 1982/85, the price of gas, whose production is closely linked to that of oil, will rocket. The cost of electricity generated from oil will also rocket. At the same time the demand, and therefore the price, of coal also will increase. That there will be a world oil shortage is now beyond doubt. It is the consequences of such a shortage that have yet to be fully recognized. A detached look at out Western lifestyle and how it is utterly dependent upon oil is a shattering experience. Even coal is transported by oil.

This is all leading up to my main point: Is wind power electricity economically competitive with power company prices, when fuel inflation is taken into consideration? The answer, particularly in good wind-speed areas, is yes. This is certainly so for large wind systems used for heating only, or for systems using a Gemini inverter. Low-output wind systems with batteries and inverter do have a much longer payback period, but there again their capital costs are relatively low.

Some assumptions about future inflation of fuel costs must be made in assessing the economic case for wind power. Take an example: Suppose your fuel bill for heating is now $560 per year. If there is no fuel inflation, you will still be paying $560 in 20 years, and during that period will have spent $11,200 on fuel. But if, as has been happening over the past five years, the cost of fuel increased 25 percent annually, your annual fuel bill in 20 years would be a staggering $43,800. What's more, your cumulative cost — the total spent on fuel over the 20-year period — would be around $191,800.

Now let us take an example of, say, a 10 kw Elektro wind generator, used for heating purposes only. The cost in Switzerland today is about $5,600, including the control gear. Add a liberal $3,500 for importation, tower and erection, and that gives a total cost of $9,100. The power could be fed into block storage radiators, or through immersion heaters to a wet radiator system, or into under-floor heating. The wind generator, costing $9,100, will produce 15,000 to 30,000 kwh per annum in good wind speed areas. A well-insulated house of between 1500 and 2000 sq. ft. in the latitude of 50° uses between 25,000 and 45,000 kwh per annum for space and water heating. The current cost of solid fuel, oil and off-peak electricity is just over 2¢ kwh, and over 4¢ kwh for daytime electric heat. Calculated here is the time in years it would take to reach break even point on this $9,100 investment in an Elektro, or indeed in any other wind-generator, given various rates of fuel inflation.

Time in Years to Break Even
(assuming a $9,100 investment in a wind-to-heat system)

Annual fuel inflation	15,000 kwh		30,000 kwh	
	2¢ kwh	4¢ kwh	2¢ kwh	4¢ kwh
25%	11	9	9	6
20%	13	10	10	7
15%	15	11	11	7
10%	18	12	12	8
5%	24	15	15	9

Even if the Elektro cost were doubled it would still be, in many cases, an attractive investment. Try it out with a pocket calculator and see. Only in the worst case, 24 years to break-even, is there any doubt. In all the other cases, an investment in wind power is justifiable.

Those who say that fuel inflation will never again be as bad as it has been over the past few years are entitled to their point of view, but it is the concensus among many who have made special studies of energy that after 1979/80, fuel prices again will be subject to rapid inflation. True, we may have a period of calm until then, but in the long run that is of little importance.

Depreciation. The life of a windmill generator is always a debatable point. On one hand, it can be said with reasonable confidence that a wind machine that is well built, sited above any turbulence, regularly serviced, thoroughly overhauled once every five years, and in general well cared for, will have a life of at least 20 years, and possibly 40 years or longer. There are many fine examples of Jacobs direct-drive generators still operating after 40 years. Direct-drive Elektros have operated continuously up in the Swiss Alps for over

30 years. On the other hand, a machine poorly made and uncared for may only last a few short years.

Low-Voltage DC Systems

This type of installation is the favorite of the self-sufficient and conservation-minded, where wood or solar energy is used for heating. Direct-heat wind systems, as discussed above, have a high capital cost but are low on maintenance (assuming the owner does his own maintenance). Wood burning, however, requires little capital outlay and high maintenance effort.

Small DC wind systems are relatively inexpensive and require little maintenance. Such a system might use the Winco 200-watt, charging one or two heavy-duty batteries to power low-voltage appliances. The cost would be approximately as follows:

Winco Wincharger 12 V 200 W with 10-ft. tower	$450
2 heavy-duty 125 amp. hr. batteries	140
Voltage regulator	70
Total	$660

That is the cost for the basic system, but rarely will the 10-foot tower be sufficient. One inexpensive way to increase the tower height is to add sections of 2-inch diameter guyed steel pipe, preferably galvanized. Added to that is the cost of 12 V cable, and if the Winco is any distance away from the house, I suggest buying a 24 V or 36 V generator, which will require less expensive cable. The usable output from such a system, in an average wind speed of 12 mph, is about 26 kwh per month, or 860 watt hours per day. Its not a lot, but is sufficient to power several 12 V fluorescent or car bulbs, a radio and possibly a small black and white television. The cost *per kwh* of such a system is high — very high — but still it has many advantages over a diesel generator. I know of a well-cared-for Winco 200 watt installation now entering its 25th year of service — and another unattended one which fell apart in less than five years.

There is a large variety of wind generators with outputs higher than the Winco yet eminently suited to power low-energy systems. Among them are the Aero Power 1 kw, Sencenbaugh 500 w and 1 kw, Dunlite 2 kw, Elektro 600w and 1.2 kw, plus many reconditioned machines such as the Jacobs 1.8 kw — all costing between $2000 and $4000 for the wind generator alone. Further details and prices are given under *Manufacturers*.

All these generators give a considerably higher output than the Winco, and therefore will give a proportionately greater degree of home comfort. Though costs vary considerably, the following gives an indication:

Sencenbaugh 1000-14 wind generator, complete with control panel	$2,650
70-foot Rohn guyed tower with top section	820
900 amp.-hr. 12 V battery bank	1,200
Total	$4,670

In an average wind speed of 12 mph, the usable power of this unit will be between 130 and 170 kwh per month. Again, when compared to current power company prices, this works out to be expensive per kwh. If amortized over 20 years the cost will be about 15¢ kwh. But then again I believe that by 1997 the power company cost per kwh will be way in *excess* of that, and of course 15¢ kwh is still very competitive with diesel generator costs, even today.

Gemini Inverter Systems

The Gemini or synchronous inverter can be scaled up more or less to any size. But here we will take an example using a fully rebuilt Jacobs wind generator:

Jacobs 2 kw	$3,500
Self-supporting tower, second-hand, 60-foot	500
Gemini inverter	1,250
Installation and incidental costs	650
Total	$5,900

This system will supply about 350 kwh in an average wind speed of 12 mph (about 250 kwh at 10 mph). Amortized over 20 years, and assuming that there are no charging differentials with the power company, this works out at the cheaper cost of 7¢ per kwh.

Large AC Domestic Systems

These wind systems allow users to take advantage of domestic equipment such as freezers, large refrigerators, color televisions, pumps, etc. The main element which differentiates them from the low-energy DC system is the inverter. This is an expensive piece of equipment, but it allows the use of all standard appliances. The cost of such an installation can be broken down as follows:

Elektro 6 kw wind generator	$5,000
50-foot Rohn guyed tower, including top section	675
Battery bank, 450 amp.-hr. 115 V DC	3,000
Delatron Inverter (see *Manufacturers*) 115 V DC to 115 V AC rated at 6 kw, (sine wave output)	4,000
Stand-by generator 2 kw	400
Incidental expenses	1,000
Total	$14,075

As shown here, the 6 kw inverter is almost as expensive as the wind generator! Smaller inverters of similar quality are only marginally less expensive. It is because of these high costs that I urge anyone considering the purchase of a wind generator to reassess his energy needs — waste is expensive and much energy is needlessly wasted. The estimated monthly output using this system in an area with an average wind speed of 12 mph is between 400 and 620 kwh. Taking an average of 500 kwh per month, the cost per kwh amortized over twenty years is 11½¢.

What follows is a further description of a number of currently installed wind generators and wind energy systems, which demonstrate the wide variety of uses to which wind power may be put.

WIND ENERGY SYSTEMS

Owner-Built Cretan
Sail Wind Generator

The 15-ft. diameter sailcloth mill shown in photo 34 generates 500 watts in an 11 mph wind. The maximum generator output is 1,400 watts and the average daily output is 6 kwh. All the power generated by this windmill is fed directly into two 700-watt 24 V DC immersion heaters. A Gemini inverter system may be added later, since the house is connected to a power company line.

The rotor hub is a sandwich of half-inch thick marine plywood, and this holds secure the 2-inch-square Columbian pine spokes which in turn support the sailcloth. The windmill is designed to feather at a speed of 100 rpm.

A specially designed gearbox with a ratio of 16:1 was built to take the speed up to 1,500 rpm, required by the motor vehicle alternator that is used. At top speed the alternator will give an output of 60 amps at 24 volts (1.44 kw). The heavy shaft and manual brake, which came from an auto junk yard, are intended to incorporate a safety device which will turn the tail vane into the same plane as the rotor when the wind gets too high — that is, over 30 mph.

The mast, bought second-hand, is a 40-ft.-high lamp post purchased from the local municipal government for a nominal sum. Its hinged base is encased in a concrete cube that is 3 ft. in all dimensions. The hinge is to allow the mill to be winched up and down for servicing. At equal spacings around a 30-ft.-diameter circle anchorage points were dug. Inverted "V's" were made out of half-

Photo 34: Sailcloth wind generator.

inch steel rod and both end sections were buried in concrete, leaving small metal arches above ground. Stainless steel yacht rigging cable was used to secure the mast. (Sailing equipment retailers can be a useful source for windmill builders.)

The great and single disadvantage of this mill is that its sails need to be hand-reefed in high winds. It can be unpleasant winching down, or climbing the tower in winter.

The house in question is an interesting one. (See photo 35.) The whole of the south-facing roof is an open channel trickle-type solar collector. Water flows from the top, down corrugated aluminum channels. This supplies summer hot water and contributes, albeit by a rather complicated heat pump process, to the annual winter space heating load. Added to this is double glazing and a high degree of insulation. The floor is insulated with 2 inches of water-proof expanded polystyrene, the walls with 4 inches and the roof 6 inches of fiberglass.

As well as being enjoyable places, the conservatory and garden produce all the fruit and vegetables required by a family of four. A great deal of hard work went into making this house-for-the-future a working reality.

Photo 35: The house with exterior insulation, a solar greenhouse, and solar roof.

Low-Energy Winco System

The small four-room cottage shown in photograph 36 is supplied with electricity from a Winco Wincharger rated at 200 watts in a 23 mph wind. The cottage lies well protected from the wind in a narrow valley, while the Winco is sited 450 feet away on the crest of a hill. It must be admitted that the sound from it reminded me of an old Dakota preparing to take off, but maybe I'm exaggerating. It is not a bad sound but it is not one I would like to have on my roof. The Winco appears to be the noisiest windmill on the market, most bigger ones making a pleasant swishing sound.

The air brake, composed of two flaps, opens to prevent over-speed by "spilling" the wind away from the propeller blades. One thing I found disquieting about this Winco was the way it rocked on its main bearing in gusts of wind. It should be said, too, that the carbon generator brushes and copper collector ring brushes both need replacing at least once a year. In contrast generator brushes on the old Jacobs mills used to last fifteen years or more.

Photo 36: The cottage with solar panels and woodburning stove.

Since the wind generator had to be placed so far away from the house, a 24-volt model was chosen. (Cable resistance at 24 volts is far less than at 12 volts, and moreover that required for a 24-volt current is much thinner than that for 12 volts and therefore is far less expensive.) All the current is taken to two 24-volt batteries connected in parallel to provide a combined storage capacity of 160 ampere hours. Power for up to seven 20-watt incandescent motor vehicle light bulbs is drawn straight from the batteries. These bulbs cast a soft, warm light and are found suitable for all activities.

It is worth noting that these car bulbs rarely fail, even when subjected to over-voltage. It is true that they do not cast as much light as fluorescent tubes, but their light is not as harsh or glaring. They can easily be focused into a beam and directed for "appropriate lighting," and there is little point in saving energy by using fluorescent if it means lighting up the whole room when only one small area needs light. Fluorescent is fine for corridors, passages and bathrooms, but have you ever tried to relax in a room lit with a buzzing fluorescent tube?

On its exposed site this Winco generates a little over 1 kwh a day — more in winter. Most of the energy is used for lighting, and the remainder is passed through a high-quality sine wave inverter to power a 6-watt radio and a 50-watt B/W television. Both radio and television are transistorized and operate just as well as their

Photo 37: Winco Wincharger high on a hill.

more power-hungry counterparts. The Winco just about provides enough power for all these needs. There is no back-up generator. If the batteries get low, out come the candles and the guitar.

The cottage, which houses two, is heated by a woodburning Jøtul stove with a maximum heat output of 6 kw. Solar panels on the roof supply all the hot water required during summer, and bottled gas takes care of cooking and winter water heating. With a slightly larger wind generator, one rated at 500 to 1,000 watts, a significant contribution could be made to winter water heating. In such a case, current for water heating would be fed directly to an immersion heater and thus by-pass the batteries. However, just as it is, the cottage is a pleasant place to live in.

Jacobs 2 kw Installation

The installation shown on pp. 63–64 is located at an altitude of 1,700 ft. on a shoulder of Prickly Mountain in northern Vermont. It provides power for the home of Donald Mayer of North Wind Power Company (see *Manufacturers*), and is located in a neighborhood that supports considerable experimentation in the field of alternative energy. Two other wind plants dominate the approach road and three nearby homes utilize solar heating systems.

The 60-foot tall tower is located some 30 feet above and behind the house, on a granite outcropping and is bolted into the ledge. The generator is a rebuilt Jacobs direct-drive model, giving an output of 2 kw (110 volts) at a wind speed of 20 mph. The blades were made by North Wind of aircraft-quality Sitka spruce, finished with a thin layer of fiberglass and painted in white epoxy. The airfoil design is an improvement upon that used by the original Jacobs.

The cost of the generator, governor and blades was $3,500. The tower, a self-supporting model, cost $500, and the Gemini synchronous inverter cost $1,250. (For further details on the Inverter, see *Windworks* in Manufacturers listings.) Installation, calculated at 12 man-days at $50 per day, cost $600. The total system, including miscellaneous parts, cost $5,900. A zoning variance was required due to the tower's height, but it was readily obtained without objection by neighbors (photo 38).

The wind plant provides 200 to 300 kilowatt hours per month, or about half of the household's electrical consumption. In times of high wind speed and low electrical consumption, excess output flows to the local utility company line via the Gemini synchronous inverter, a solid state unit which forms the interface between the public power grid, the generator and the house wiring. When the wind fails, current is provided from the power company to the house through a standard company meter in the usual way.

The system is similar to that used for many years in regenerative drives such as elevators, decellerating electric trains and other heavy equipment, and its advantage is twofold: the inverter unit converts the wind generator's variable DC output to constant-voltage AC, and it synchronizes the signal with the power company's sine-wave. In addition, the wind system's overall cost is cut in half by elimination of battery storage and conventional inverter. In this case, the installation has been extensively metered by the local utility (Green Mountain Power Company) to help them determine an equitable rate structure for customers like the Mayers who do not utilize the power grid as a primary electrical source.

The Mayer family had several reasons for wanting to use wind-electrics as a source of energy for their home. As an expression of their philosophy of decentralization, the system retains the option of becoming fully independent at some future date with the addition of battery storage. The system increases awareness of energy consumption through its visual impact. Thus it is also an educational tool in the community, demonstrating wind power's validity as an energy source.

The other energy systems in the house are compatible with this philosophy. Cooking takes place on an antique wood-fired cookstove, a modern propane stove serving as a backup in hot weather. A Clivus Multrum composting toilet receives all human waste and kitchen scraps. Requiring no external energy flush input, the Clivus will eventually produce a high-quality garden fertilizer. Through use of the Clivus, the three Mayer children, who range from three to ten years of age, have learned to identify biodegradable materials in their environment.

Space-heating is a major use of energy in this northern climate, and the Mayer home utilizes passive solar heating techniques to provide for a substantial portion of the heating needs. Virtually

Photo 38: 2 kw Jacobs Wind Electric.

the entire south face of the structure consists of a 45° pitched roof glazed with acrylic sheeting. The floor of the family living spaces consists of a heavy concrete slab faced with slate. During periods of direct or reflected solar radiation the floor acts as a heat sink, aided in its storage capabilities by a massive central chimney (photo 39).

When the Mayer house is complete, panels of insulating material

Photo 39: The Mayers's natural energy home.

will be available to slide into place behind the glazing, retarding re-radiation of stored heat from the slab into the night sky. The panels will be stored behind a bank of liquid-type solar collectors, that provide most of the family's hot water needs. Supplimentary heating during cloudy weather is provided by a space heater fueled with hardwood culled from the surrounding forest. A propane central furnace allows the family to be absent for several days at a time without fear of frozen plumbing. Overall heat requirements are reduced by the three inches of urethane foam insulation in the building's exterior walls.

A clearing just in front of the house allows space for several bee-hives, a vegetable garden and pasture for two work horses. These elements compliment the Mayers' decentralist philosophy: developing the skills and tools necessary for achieving a high degree of material self-sufficiency.

Hydrowind

The Hydrowind, erected in September 1976, was the first wind generator to utilize a hydraulic system. It was designed by Merrill Hall and Vince Dempsey for use by the New Alchemy Institute at their "Ark" research center on Prince Edward Island, Canada. P.E.I. leads North America in its laudable decision not to use nuclear energy, but instead to take a more positive role in utilizing a coal/solar/wind energy system.

The New Alchemy Institute, a non-profit organization, built the "Ark" on a coastal site of 137 acres. It is an integrated structure, comprising a residence, extensive greenhouse, fish-farming unit, research laboratory, barn and tool shed. The south facade of the building is pure solar, with a vast expanse of solar collector and greenhouse glazing. Rocks and water in the building store up to 13.2 million BUT's of solar heat, which, when combined with a wood-stove, will heat the "Ark" for all December, even if the sun never shines! From the residence, there is a spectacular view of the sea. The afternoon sun shines freely into the living room, dining room and bedrooms (photo 40).

Photo 40:   The Ark on Prince Edward Island, Canada.

The Hydrowind, the first of four such machines to be used at this site, generates a maximum of 7 kw in a 25 mph wind speed. The three-bladed upwind propeller has a diameter of 20 feet. The blades utilize a lightweight design based on an internal tension system and are covered with an aluminum skin. Hydraulic governors are used to vary the pitch of the blades, thus giving a constant propeller speed in varying winds. A 3:1 belt drive connects the shaft to the hydraulic transmission, which in turn drives the alternator on a platform half way up the self-supporting tower. The overall efficiency of the hydraulic system is about 90 percent (photo 41). (It is interesting that a company in England (see the "Elteeco System") was working on a very similar system at the same time, each firm unaware of the activities of the other.) A Gemini synchronous inverter is used with the local power company to provide alternating current at 110 volts, 60 Hz.

The Hydrowind, still under test, is not yet commercially available, although the design is such that it is suitable for regional manufacture. The performance of the energy systems in the "Ark" during the severe winter of 1976/77, was even better than expected. The size of the "Ark" probably represents the minimum climatic mass required for its operation as an economic unit with part-time tending. It has the capacity to raise up to 20,000 fish at one time, 10,000 cuttings of valuable fruit and nut trees, plus the ability to raise year-round greenhouse foods and flowers within its integrated design.

Photo 41: The New Alchemists's Hydrowind machine.

The goal of the New Alchemists is to show how to build living structures that pay for themselves. Their excellent *Journal* and membership of the Institute are available from: New Alchemy Institute, P.O. Box 432, Woods Hole, Massachusetts 02543.

The Elteeco System

This novel windmill system, developed in England, has many attractions: simple and sturdy blade configuration, hydraulic "gearing" and transmission, two ground-level generators (one an induction type for operation with a local power line to provide utility-type electricity and the second an alternator for resistance heating), and is aesthetically pleasing, being reminiscent of old-fashioned mills.

The Elteeco wind generator, opposite, the first of its kind, is located near the home of the inventor, Sir Henry Lawson-Tancred. Rated at 30 kw in a 20 mph wind, it is principally designed to heat and light large country homes, although it can be used for many other purposes, such as greenhouse heating, etc.

The Elteeco uses a three-bladed, fixed-pitch propeller made of steel spars supporting fiberglass envelopes molded in an aerodynamic profile. The blade structure is strengthened in a basic and elegant way to withstand wind speeds of up to 120 mph. Guy wires connecting the center of each blade can be seen in the photograph, as can the metal supports which connect to a central tripod.

The 60-foot diameter propeller is mounted on a horizontal shaft, which in turn is supported by a 45-foot high, four-post, timber-covered tower. The propeller shaft is not strictly horizontal but is set at an angle of 5° so that additional clearance is provided between the blades and the tower base. The tail vane directs the blades into the wind.

The shaft is connected to four hydraulic gear pumps by a speed-increasing gearbox. As a result oil, under pressure, is passed to the "energy integrator," which in turn supplies hydraulic oil under constant pressure through shut-off control valves to two hydraulic motors, of low and high power respectively. The low-power motor operates a 5 kw induction generator in phase and voltage alignment with the power company supply from which it is energized. The high-powered motor drives a 25 kw alternator, separate from the power line, and whose output is used for heating purposes only.

Full system output of 30 kw is reached in a 20 mph wind, although the average wind speed on this test site is 15 mph. The expected output per year is 90,000 kwh, 15 percent of which will be power line-type electricity from the induction generator. The remainder, from the alternator, will go to space heating.

Excess energy in higher wind speeds — up to 27 mph — is dissipated through a pressure-reducing valve. A further rise in wind

Photo 42: The Elteeco three-blade windmill.

speed above 27 mph causes centrifugally operated blade flaps to extend, thus providing an air brake on the propeller. The Elteeco System is further equipped with a wind-speed sensor connected to an electronic circuit. In the event of wind speeds above 35-40 mph, spring-operated (but electrically retained) band brakes stop the propeller — and there are further safety devices attached to both generators. In the several months of testing so far, the system has withstood winds of up to 90 mph.

Elteeco offers three wind systems: the first as outlined above, which is still under test; the second drives a 30 kw induction generator in parallel with a local power line, assuming the utility is agreeable to the arrangement; the third drives an ungoverned alternator suitable for heating purposes only, this producing 50 kw in a 25 mph wind. At the time of writing, one type 3 (50 kw) is for sale at $19,200. Prices for the other two types are not yet available. It will be interesting to see how the Elteeco Wind Machine proves itself over the next year or two.

Zephyr Wind Installation

The building shown in photograph 43 was designed as an "all-electric house" back in 1970, when electricity cost 1.7 cents kwh. Within the space of a few short years the cost had risen to 4.3 cents kwh and, as there seemed no end to these increases, the owner decided to invest in an inflation-free wind generator. Today, with the help of two woodstoves, the Zephyr Wind Dynamo supplies all space and water heating. Utility electricity is relegated to lighting, motors and backup.

The Zephyr is a three-bladed downwind model that gives a maximum output of 15 kw in a 30 mph wind. The expected output on this site, a rocky shorefront with an average wind speed of about 15 mph, is about 20,000 kwh yearly. The electrical output has a varying voltage and frequency, but since all the output is used for resistance heating, batteries and inverter are unnecessary. These components can be added later if an independent source of utility-type electricity is required.

Photo 43: The Zephyr Wind Dynamo.

The Zephyr is wired to supply two separate 3-phase circuits of 7.5 kw each. Careful resistance matching is required to achieve best efficiency, and the system powers four appliances, which can be run selectively. First is a space heater with nine 1 kw incandescent lightbulbs. The second consists of 70 building bricks with nichrome wires threaded through the cores, making up a 7.6 kw storage heater. Third is a waterbox, located in the master bedroom, which heats the room as well as incoming cold water before it enters the hot water tank. The waterbox is valved so that in windy hours water may bypass the commercial electric heater for direct use. The fourth appliance, a 15 kw immersion heater, is used only in summer for heating the swimming pool. A portion of the remaining heat goes to a south-facing 200-square-foot growing room, which is maintained at 50° at night. This glass-walled room accounts for the use of more than 500 kwh per month of electricity in mid-winter.

The total cost of duplicating this whole installation would be approximately $20,000 — expensive, and justifiable only on the grounds that fuel prices will continue to increase as they have over the past few years. The great benefit of this resistance heating system is that it involves no batteries or inverter.

This just leaves the difficult question of how long the Zephyr wind generator will last. I cannot comment on the plastic composite blades, except to say that with a good protective coating I hope they would last for "many a long year." The direct-drive feature means that there will be no future gear-box problems. The six-pole, permanent magnet alternator is of solid, low speed design and is well protected from vibration by rubber mountings. The bearings will need replacing once every seven years or so, but at two dollars a time that won't be much of a problem.

The owners of this installation, Giff and Lois Horton are skilled and energetic people, who, with the help of their sons, designed and built this innovative and energy-conserving house long before it became fashionable. They are excellent example customers for wind generators, and in particular for field-testing new machines like the Zephyr, since they recognize the experimental nature of the design and are prepared to deal with the many annoyances and difficulties inherent in de-bugging a new product. This ability to grapple head-on with the technical, financial, legal and institutional problems raised by the introduction of a new natural energy device makes the Hortons and others like them the "new pioneers."

For further details of the Zephyr Wind Dynamo see *Manufacturers.*

Conservation House

The house shown in photograph 44, possibly the most advanced of its kind, depends for its energy on a 2 kw Dunlite wind generator which provides about 15 kwh of electricity a day. This meager energy output, scarcely one-fifth of the average home use, however is capable of providing all normal domestic requirements — temperature control, cooking, water heating, lighting and small power.

The three-bedroom house is heavily wrapped in 18-inch thick fiberglass cavity insulation! This gives a "U" value of 0.0737 and effectively reduces the energy required for winter heating to one eighth of normal. The windows are *double* double glazed — too bad if a baseball goes through one!

Temperature is controlled by a small heat pump driven by a 150-watt motor. The heat pump, which can be used to heat or cool, has an optimum coefficient of performance of 3:1. This means that for every unit of energy input there is an output of three units of heat (see Air Distribution diagram, p. 74). Air flow within the house is carefully controlled to avoid heat loss.

Photo 44: "Conservation House" with 2 kw Dunlite. Output from the Dunlite would be considerably improved by increasing the tower height. Fortunately, the prevailing wind comes across open ground.

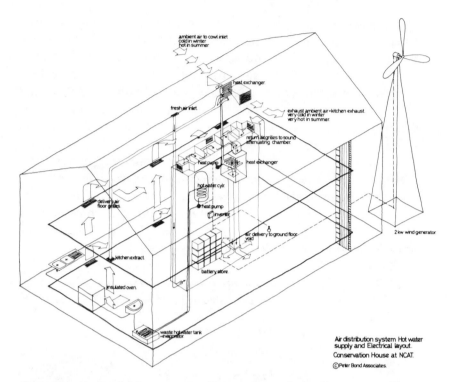

Figure 45. Air distribution system and hot water supply for Conservation House. (Copyright by Peter Bond Associates.)

Domestic hot water also is provided by a small heat pump, this time with a C.O.P. of 2:1. Significant heat for water heating is ingeniously extracted from heat in the waste water. Instead of pouring it down the drain, waste hot water goes to a tank under the ground floor where all the heat is extracted (gremlins at work!) before being sent on its way cold. A shower is used instead of a bathtub and hot water is in general used with care. Rain water is collected, filtered and used for all domestic purposes except drinking.

An electric cooker, operating on 110 volts DC, does all the cooking on approximately half the energy normally required, by means of a patented design which retains the heat in the oven, which also is enclosed in six inches of insulation.

Lighting in the Conservation House is provided by low-wattage fluorescent fittings. Power socket outlets are provided as normal for domestic appliances.

Power from the wind generator is fed to a 182 ampere hour, 110 volt battery bank (see Electrical Layout diagram, Figure 46.) A small standby generator is available for any long, windless periods and also to allow for periodic maintenance on the Dunlite.

Conservation House was designed for a site at latitude 52° but its principles could apply to a much wider spectrum of climates. Details aside, and much as I appreciate the house, I don't feel I would be happy living there. Inside I find it reminds me of an isolation ward in that one is cut off from the outside world. Bird song, rain, or wind in the trees cannot be heard through the layers of insulation, and moreover, the small windows do not open. But not everyone agrees with my criticism, least of all those who live in the house. For further details on the Dunlite see *Manufacturers*.

Figure 46. Electrical layout of Conservation House.

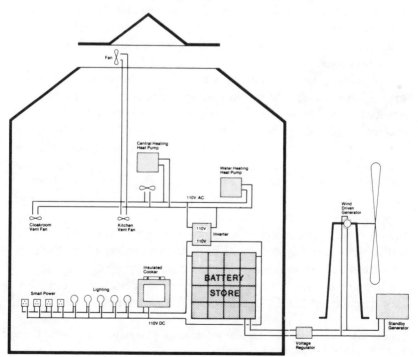

A Wind-Powered
Solar Greenhouse

As the demand for oil exceeds supply, in just a few short years' time we are going to have to reassess not just our domestic energy requirements, but also our food requirements and their sources. The oil-intensive agri-business which currently provides our food has a short life ahead. A close look at modern agriculture and our food distribution network will show how totally oil-dependent it is. Anybody who does consider this dependence will, almost certainly, want to make plans to provide food for his or her own family.

The system shown in photographs 47 and 48, a wind generator and greenhouse, is most interesting. When the wind blows, greenhouses more rapidly lose their heat, a problem that can be solved by insulating the greenhouse (especially the north side and at nightfall), and then adding a simple wind energy system to produce heat, replacing that which is being lost at the same time. In such a system there is no need for batteries or inverter, since all the current is simply fed into an immersion heater, similar to the heat element in a water tank.

The wind generator shown is not strictly essential because the greenhouse (originally developed at Helion) is of a particularly efficient design and so requires little external energy. Double glazing and insulated walls help to retain heat during cold periods. A curved rear (north) wall acts as a reflector to provide increased light and heat to the interior during the winter months with low sun angles and short days. But this is an exceptional greenhouse of recent design* and does not solve the problem of what is to be done with existing greenhouses. Insulation will help greatly, but additional power must be provided. In many cases wind power, with or without a heat pump system, will be sufficient to maintain the right conditions. A UK company, Wind Energy Supply Co. (see *Manufacturers*) has done considerable pioneering work in this direction. In particular, WESCO has developed a 60-foot diameter 100 kw

---

* Home-construction plans for this solar greenhouse are available from Provider Greenhouse, Box 49708, Los Angeles, CA 90049.

Photos 47 & 48: Exterior and interior views of a wind-powered solar greenhouse.

wind turbine, intended for use mainly with commercial green-houses.

The smaller domestic system shown above features the Helion 12-foot diameter downwind turbine. Built by the owner from the Helion 12/16 Construction Plans, (see *Manufacturers*), it provides about 1,500 kwh of heat per annum on this particular site with its average wind speed of 10 mph. The total cost of the wind generator was $1,000, and its construction took about ten man days of concentrated though enjoyable work.

An Owner-Built
Multi-Blade Windmill

Arnold Stead is a retired chimney sweep who enjoys playing the double bass and watching the stars through his telescope. The small house where he and his wife live is heated by a wood stove, its fuel cut by a circular saw driven by a gaily painted multi-blade windmill. The mill has worked happily and trouble-free for the past 15 years.

About 20 years ago Stead wondered if he could put some of the waste he saw around the countryside to useful purposes. Even though he had no previous mechanical training or experience, he decided to build a windmill. The blades and shaft of the first mill he built were not strong enough, vibration set in and a 60 mph wind blew it to pieces. But that was years ago, and he has learned much since. If that hadn't happened, his second mill would not be so good.

Not to be beaten, Stead set about building another mill, and the results of his efforts can be seen in photograph 49. The 8½-foot diameter multi-vane propeller, which also has eight small blades at the hub, spins in the slightest breeze and looks like a colorful, revolving marigold. The propeller alone weighs a massive 225 pounds and is built from scrap, as in fact is everything else on this mill: old bed parts, second-hand pulleys. Even old railway tracks are used to bear the weight of the windmill on top of the workshop (photo 50). If this sounds all very slapdash, it isn't, for it all fits and works. At the other end of the propeller shaft is a 168-

Photo 49: Arnold Stead and his wind machine.

Photos 50 & 51: Stead's "Pennine Wind Engine" (left), and its wind-driven circular saw (right).

pound flywheel, again scrap, and it took Stead six months before he came across the right one.

The mill's main bearing is built to withstand the pressure of five tons. A crown wheel and pinion takes the drive straight down into the workshop, where it operates two grindstones (for sharpening tools and knives) and the saw. Exact performance figures are not available, but it suffices to say that the mill does its job, producing up to 5 hp (3.8 kw) in a 30 mph wind. It is capable of sawing fair-sized logs (photo 51), and moreover has stood the tests of time and wind speeds of 90 to 100 mph!

The secret of Stead's success is the way in which he built the mill. Every night, before starting the project, Stead would go to bed and, instead of going to sleep, he would shut his eyes for about one hour and see the mill to its finish. He visualized the mill, complete and working, before starting to build it. Then, knowing exactly what parts he wanted, he would go about his business as usual but keep an eye out for the necessary ingredients. If this meant waiting for months, well, he could wait.

This is somewhat the way millwrights of old used to build the Dutch four-arm mills. When visiting the site of a proposed windmill for the first time the millwright invariably would infuriate his employer by standing for hours and even days just looking at the site and apparently not doing any work at all. Probably he spent this time looking for the best exact site and visualizing how he would build the mill. Finally, and in his own good time, the millwright would start the construction work.

I am convinced that the more creative and careful thought put into a system or machine before its construction, the longer its life will be and the less trouble it will give. The rapid obsolescence and shoddiness of many of our industrial products is witness to the lack of careful consideration built into the original idea. By contrast, Arnold Stead's mill is endowed with an almost life-like quality.

"The material *needs* of human beings are limited and in fact quite modest, even though our material *wants* know no bounds."

Small is Beautiful, by Schumacher

MANUFACTURERS
&
RESTORERS

When writing to any of the companies listed below, please enclose a dollar or two to help defray printing and mailing costs, which can become a crippling expense to a small firm.

Aero Power
2398 4th Street
Berkeley, CA 94710
Phone: (415) 848-2710

Aero Power has been in the wind business for about six years, during that time having manufactured about 700 sets of propeller blades for a wide variety of reconditioned machines and for experimental college applications.

Aero Power also manufactured a "Model A" twin-bladed wind generator rated at 1,000 watts in a 32 mph wind. This has now been replaced by a new three-bladed "Model A" rated at 1,000 watts at 25 mph.

Aero Power's "Model A" 1 kw wind generator.

The propeller has a diameter of 8½ ft., its variable-pitch blades made of Sitka spruce. Transmission to the alternator is by means of a helical gear with a ratio of 2.5 to 1. The alternator starts charging at 42 watts in a 7 mph wind and reaches a maximum output of 1,050 watts at 25 mph. At high wind speeds the blades automatically feather to prevent damage to the unit. The price of the Aero Power with control box is $1,850.

Aero Power also sells Rohn towers, inverters and batteries (send $1.00 for their catalog).

The following is an example taken from their range:

60-ft. guyed tower	$605
500-watt square wave inverter	$350
12-volt 230 amp-hour battery set	$260
Aero Power wind generator	$1,850
	$3,065

Aerowatt
37, Rue Chanzy — 75011
Paris, France

Aerowatt manufactures the most expensive range of wind generators available, and one reason is that their machines are rated at very low wind speeds. For example, their 4 kw mill gives its full output in a 15 mph wind and will cost 77,000 French francs, their 1 kw mill 35,200 F.F. The other reason for their cost is that they are also built to operate in very severe wind speeds — up to 175 mph.

Pennwalt Automatic Power, P.O. Box 18738, Houston, Texas 77023, is U.S. distributor for Aerowatt. The 4 kw costs $34,350, and the 1 kw $15,475. At these prices they are of limited interest to domestic users.

American Energy Alternatives (Amernalt)
P.O. Box 905
Boulder, CO 80302
Phone: (303) 447-0820

The Amernalt wind generator.

Amernalt currently offers two wind generators, its 8-ft. diameter horizontal-axis rotor similar to the old multi-blade mill. One machine is rated at 1.5 kw in a 25 mph wind and costs $3,450. The second, rated at 2.5 kw at 40 mph, costs $3,650. A 12-ft. diameter rotor will be available shortly, generating 2.5 kw at 24 mph.

Both machines have withstood winds in excess of 90 mph. The aluminum blades are fixed to steel spokes. The unit is fitted with an automatic brake, and comes complete with electrical overload protection, voltage regulator, etc.

American Wind Turbine
1016 East Airport Road
Stillwater, OK 74074
Phone: (405) 377-5333

This company manufactures a multi-blade rotor similar to the Amernalt. Their 16-ft. wind generator reaches its maximum output of 2 kw at 20 mph (1 kw at 15 mph). Prices are put tentatively as follows:

Wind turbines	$914
16-ft. tower	$750
Mounting kit	$125
Alternator & Controls	$625
	$2,414

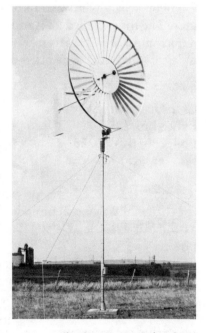

An American wind turbine.

AWT also manufactures 12- and 16-ft. turbines suitable for heating, powering electric motors or electric water pumps, their prices similar to above. The use of a switch box ($65) will allow one turbine to serve all three purposes.

Ampair Products
Aston House
Blackheat
Guildford
Surry, England

Ampair manufactures a 50-watt wind generator intended for marine applications.

Comcop
Box 1267
Redwood City, CA 94064

Comcop sells a wind generator kit that costs $250, less the generator. The three-bladed 12-ft. diameter aluminum propeller generates a maximum of 500 watts in a 20 mph wind when fitted with a suitable car alternator. The plans cost $5 and a catalog $1.

Coulson Wind Electric
RFD 1 Box 225
Polk City, IO 50226
Phone: (515) 984-6038

Roland Coulson sells a wide range of reconditioned wind generators originally manufactured back in the Thirties and Forties, such as the Air Electric, Delco, Parris-Dunn, Wincharger and Windpower.

The Air Electric Company produced two 32-volt machines, 2 kw and 3 kw. On both models the generator cowling and tail were one continuous piece with vents for air-cooling the generator. The two-bladed propeller had an 80-pound flywheel for stability. Governing was achieved by means of curious looking paddle deflectors.

Coulson's 32-volt workshop (left) and its Windpower 2.5 kw generator.

Delco and Parris-Dunn both manufactured small, 200-watt wind chargers for radio operation. In high winds, the Delco tail vane would fold and thus draw the propeller out of the wind. Parris-Dunn also manufactured a 2 kw model, and had an unusual governing system in that high winds would lift the propeller upwards into a vertical plane, out of the wind.

Wincharger Corporation started in 1927 and manufactured several hundred thousand wind generators before being bought by Dyna Technology, Inc., which continues to produce the Winco 200 watt (see also page 109). As well as the 200 watt model, Wincharger also produced two- and four-bladed 650-watt and 1,200-watt models. The two-bladed machines all used centrifugal air flap governors, as does the Winco 200 watt today.

Windpower Corporation produced 1.2 and 1.8 kw three-bladed downwind machines. The 12-foot diameter propeller was directly connected to the 32-volt generator. The blades were governed by fly-balls. A manual shut-down brake was used in storm conditions. Coulson, an emporium of rebuilt machines and parts, also sells reconditioned Allied, Black Swan and Alamo Dynamo wind machines as well as new and used batteries, towers and 32-volt equipment.

Coulson powers his workshop with a 1.2 kw Wincharger on an 80-foot tower and a 2.5 kw Windpower on a 65-foot tower. All the workshop equipment is run on 32 volts DC: a lathe, valve and bench grinders, drills, air compressor, vacuum cleaner, radio and lights.

Delatron Systems Corporation
553 Lively Boulevard
Elk Grove Village, IL 60007

Delatron, which also manufactures heavy duty batteries, recently announced the availability of a new range of cost-competitive DC to AC inverters developed for wind and water power users. There are three models initially:

36 V DC to 120 V AC output, 3 kw capacity	$2,800
120 V DC to 120 V AC output, 3 kw capacity	$2,800
120 V DC to 120 V AC output, 6 kw capacity	$3,995

Write the company for further information.

Dominion Aluminium Fabricating, Ltd.
3570 Hawkerstone Road
Mississauga
Ontario L5C 2V8
Canada
Phone: (416) 275-5300

DAF manufactures vertical-axis Darrieus turbines. Both the 15- and 20-ft. models, designed to withstand gusts up to 130 mph, require motor start. The 15-ft. model is rated at 4 kw in a 23 mph wind, while the 20-ft. model generates 6 kw at the same wind speed. The choice of generator voltage on the 15-ft. model is 24

DAF twin-bladed Darrieus rotor.

or 110 V. It comes complete with electrical control gear and stub tower, and costs about $8,000.

DAF Average Monthly Output — Kilowatt Hours

Turbine Diameter	Average Monthly Windspeed									
	Miles Per Hour					Meters Per Second				
Feet–Volts	9	11	13	15	18	4	5	6	7	8
15–24V	110	190	290	420	680	110	200	310	500	670
15–110V	110	210	360	560	1000	110	220	390	640	990
20–110V	210	400	680	1070	1900	210	420	745	1040	1860

Dunlite
Pye Industries Sales Pty. Ltd.
22 Hargreaves Street
Huntingdale,
Victoria 3166
Australia

The Dunlite 2 kw wind generator has been on the market for 30 years and in that time has proved itself to be a sturdy and reliable machine. Three galvanized steel, variable-pitch blades form the standard 13-ft. diameter propeller. Centrifugal governor weights are mounted on the blades. A heavy-duty helical gear takes the maximum propeller speed of 150 rpm up to the alternator's 750 rpm. The specially built multi-pole brushless alternator generates a maximum continuous output of 2 kw at 25 mph and starts charging in 10 mph wind.

Dunlite 2 kw on a Rohn guyed tower. Erected by Enertech.

The standard propeller is designed to withstand a maximum wind speed of 80 mph, while a special short propeller 10.5 ft. in diameter, will withstand winds of up to 120 mph but will give a lower output than the larger propeller at given wind speeds. This 80 mph wind speed limit is the only objection to the Dunlite, (and I do know of one machine that was destroyed by winds in excess of 90 mph), but this is a problem only where such high wind speeds are experienced.

The cost of the unit in Australia is $2,000 (Australian dollars), but don't forget that importation will add considerably to the cost. Remember, however, that if you purchase a Dunlite from an experienced agent you will also receive the benefit of their years of experience, and that is worth a lot.

Dunlite plans to manufacture a 5 to 6 kw unit shortly, expected cost $3,000 Australian.

See page 73 for installation example of a 2 kw Dunlite.

Dwyer Instruments, Inc.
P.O. Box 373
Michigan City, IN 46360
Phone: (219) 872-9141

The Dwyer Wind Meter and the Speed Indicator (see page 10) are manufactured by this company.

Edmund Scientific Co.
222 Edscope Building
Barrington, NJ 08007
Phone: (609) 547-3488

The Edmund Wind Wizard generates 600 watts at 25 mph. The 3-bladed wooden propeller is geared to a 12 V Delco generator. It is regulated by a folding tail, "assembles in 10 minutes," and costs

Edmund Wind Wizard.

$825. It is not designed to withstand more than 50 mph winds, and so is quite useless for any serious continuous use.

Elektro G.m.b.H.
St. Gallerstrasse 27,
Winterthur, Switzerland

Elektro, a tiny outfit, has been quietly manufacturing wind generators in Switzerland for the past 37 years and all went well until about 1969 when suddenly everybody wanted to buy Elektros. Reeling under a deluge of mail and orders, the small workshop where Elektros are hand-built became hurried and cluttered. New people were employed who did not understand the fine art of wind-

Elektro 5 kw generator with gear, shown without casing.

craft, and so quality control slipped. About the same time Elektro allowed itself to be pushed into premature manufacture of a 10 kw generator by an English company. The result was that the blades on the first batch of 10 kw mills broke, and Electro's reputation suffered badly as a result.

High in the Swiss Alps, a direct-drive Elektro.

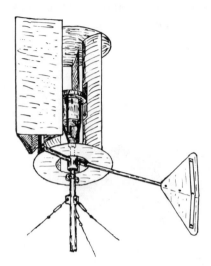 

Vertical-axis Elektro (left) shown powering a mountain rescue post in the Alps.

It must be said, however, that a company expert subsequently personally visited and de-bugged most of the faulty Elektros in Europe. Having recently met Mr. Kern and Mr. Schaufelberger of Elektro, I'm personally satisfied that they have matters back in hand again and are intent upon improving their quality control.

After all that, it should also be said the Elektro has many happy customers.

Once an Elektro is working, it tends to last a long time, 30-40 years. The following is a selection from their wide range of wind generators.

Max Output	No. of Blades	Cost
600 watts	2	S.Fr. 4,300
1,200 watts	2	S.Fr. 5,200
5/6,000 watts	3	S.Fr. 9,500
8/10,000 watts	3	S.Fr. 15,000

These prices include export packing, control panel and voltage regulator, but do not include tower, masthead, or various other accessories. Import duties will also add to the cost. It is frequently best to purchase Elektros from a local agent, provided there is one, and his experience will be helpful, also.

Elektro also sells suitable towers, batteries and inverters and manufactures wind generators for heating only. Elektro blades are made of wood, their variable pitch governed by centrifugal weights. All models up to the 5/6 kw are direct drive. Field magnet and permanent-magnet alternators are used. Write to Elektro for details, or send S.Fr. 50 for their recently written manual, "Thirty-seven Years Experience with Elektro Windmills."

Elteeco Ltd.
Aldborough Manor
Boroughbridge
Yorkshire, England

Elteeco manufactures 3-blade fixed pitch propeller systems with hydraulic linkage and drive to 25, 30 and 50 kw alternators, designed for linkage with local power company lines or for resistance heating. For more details, see page 68. Send $2 to the company for further information.

Enag S.A.
Rue de Pont-L'Abbe
Quimper,
Finistere, France

This company is reported to manufacture twin-bladed wind generators with outputs from 180 to 2,000 watts, but they seem reluctant to let anyone know about them!

Grumman Energy Systems
4175 Veterans Memorial Highway
Ronkonkoma, NY 11779
Phone: (516) 575-6205

GES, a division of the huge Grumman Corporation, manufactures the Grumman Windstream 25, a three-bladed downwind mill. The variable-pitch aluminum blades are electronically controlled, while centrifugal speed governors are located at the tip of each blade. The drive from the propeller is geared-up to the alternator. The power options are 20 kw at 28 mph or 15 kw at 26 mph. The Windstream and its concrete tower are designed to withstand 130 mph winds with a safety factor of 1.5. Automatic shut-down is at 50 mph. Costs are as follows:

Windstream 25 $20,000

complete with control panel for battery charging or for operation with 15 or 20 kw Gemini Inverter (including lightning conductor).

Steel-reinforced
45 ft. concrete tower $1,965

Grumman's first customer is the New York State Energy Research & Development Authority, which is using the Windstream to power a farm and is monitoring its performance.

Grumman Windstream 25 estimated output (based on The British Research Association statistics.)

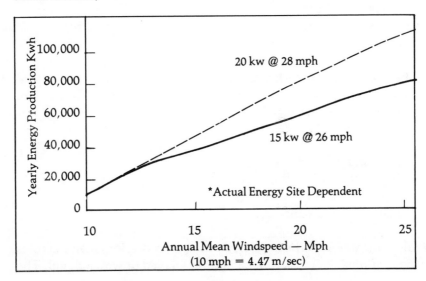

Helical Power Ltd.
36 Fontwell Drive
Glen Parva
Leicester LE2 9ML, England

The Helical rotor, as described
earlier under Windmills (see page
34), consists of two vertical blades
forming a helical twist. These
mills are not yet commercially
available, though production of
a 1 and 2 kw model is planned.

The Helical rotor.

Helion Inc.
Box 445
Brownsville, CA 95919
Phone: (916) 675-2478

Helion is a non-profit research/
education organization in renewable energy areas started many
years ago by Jack Park, who is
also the technical editor of *Wind
Power Digest* (see *Bibliography*).
Helion publishes the "12/16 Construction Plans" for the home construction of a 3-bladed downwind propeller turbine. The 12-
footer has a maximum output of

Helion wind generator.

2 kw at 25 mph. A kit based on these plans is marketed by Topanga
Power and a finished model is manufactured by Kedco, Inc., (both
listed later in this section).

Helion also manufactures a DC watt meter ($98), which is useful
for monitoring a DC wind electric system. They have also designed
an Energy Source Analyzer which measures and records speeds on

10 recorder channels, measures total solar insolation, and provides total time indication. The price for this miniature meteorological station is $1,200.

The home of Helion Inc. is a semi-self-sufficient ranch, and now the organization is involved in the design of similar projects for others.

See page 77 for a Helion installation.

Kedco Inc.
9016 Aviation Blvd.
Inglewood, CA 90301
Tel: (213) 776-6636

Kedco manufactures a family of four wind generators based on Jack Park's three-bladed down-wind plans, (see *Helion*, above). Models 1200 (12-ft. diameter) and 1600 (16-ft. diameter) are both rated at 1,200 watts maximum for battery-charging applications, while models 1210 and 1610 are supplied with 2,000 watt DC generators of permanent magnet design for synchronous inverter operation and for direct heating applications.

The fundamental difference between the 1200 and 1600 series is

Kedco model 1600.

the blade diameter. The increase in diameter from 12 to 16 feet nearly doubles the energy yield. Kedco offers for a nominal price a series of notes explaining the most appropriate ways to use their wind generator range.

All models have aluminum blades and the following features: automatic blade feathering, ground shut-off and re-set cables, automatic vibration-sensing shut-off and one year's warranty.

Prices: model 1200 $2,295   model 1600 $2,895
         model 1210 $2,595   model 1610 $3,195

For export packing add $100.

Output chart for the Kedco wind generator range.

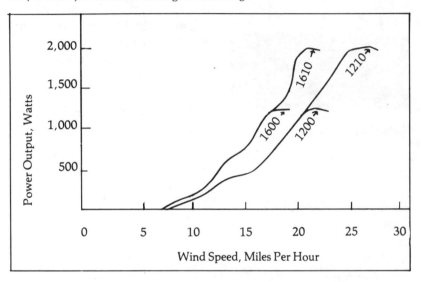

Low Energy Systems
3 Larkfield Gardens
Dublin 6,
Ireland

Low Energy Systems has developed a vertical-axis sailwing mill
suitable mainly for water-pumping and other mechanical purposes,
its simple construction and low speed output making it comparable
to the horizontal-axis Cretan mill.

The rotor consists of two or more sailwings, each formed from
a rigid spar which is positioned at the leading edge of the sail. To
this spar two or more rigid ribs are attached at right angles. The
trailing edge of the sailwing is held in tension between the ends of
the spars. Its surface is made from cloth.

Vertical axis sailwing.

In operation the sailwing takes on an airfoil shape with a concave surface facing into the wind. During one complete revolution of the rotor the sailwing switches the concave surface from one side to the other automatically.

Unlike the Darrieus rotor, this one is self-starting, and its estimated efficiency of 30 percent will shortly be tested in a wind tunnel. Home construction plans for this mill are available from the company for $3 postpaid.

Lubing-Maschinenfabrik
Ludwig Bening
2847 Barnstort
Postfach 110,
West Germany
Phone: 05442-625

Lubing produces about 50 different sizes of wind-driven water pumps and only one type of windmill generator — a downwind horizontal-axis machine. The Lubing mill, rated at 400 watts in a 27 mph wind, has been steadily improved over the past 25 years of manufacture and has achieved a reputation for robust design and reliability.

The six-bladed Lubing propeller as shown at the right has three small, fixed-pitch blades which start the mill at 9 mph. The three

Lubing wind generator.

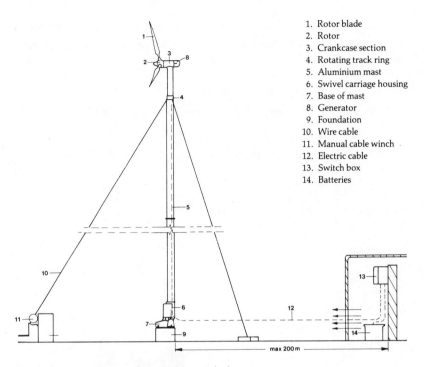

1. Rotor blade
2. Rotor
3. Crankcase section
4. Rotating track ring
5. Aluminium mast
6. Swivel carriage housing
7. Base of mast
8. Generator
9. Foundation
10. Wire cable
11. Manual cable winch
12. Electric cable
13. Switch box
14. Batteries

Lubing generator mounted on a folding tubular tower.

larger variable-pitch blades produce the bulk of the power at higher wind speeds. All the blades are made of epoxy resins reinforced with fiberglass. A centrifugal governor, fitted to each of the variable pitch blades, prevents them from exceeding the maximum speed of 600 rpm.

Transmission of power from shaft to brushless alternator is by means of a two-stage oil-bath gear with a ratio of 5.5:1.

Output from the alternator is converted from AC to 24 volts DC at the control panel. The electronic controls regulate the charging of the batteries automatically, and when a battery voltage of 28.5 volts is reached the charging current is cut off.

The price is high — 5,537 German marks (DM) for the basic unit with 3-ft. stub tower and control panel.

The Lubing is also sold complete with aluminum tubular tower in three sizes. The tower is easy to bolt in place and has the added advantage of being hinged at the base, which enables the owner to raise and lower the mill by means of the winch provided. This makes servicing and the annual gear-box oil change a relatively simple job:

Lubing wind generator
complete with 23-ft. tower   DM7481

Lubing wind generator
complete with 33-ft. tower   DM7895

Lubing wind generator
complete with 42.5-ft. tower   DM8329

Natural Power, Inc.
Francestown Turnpike
New Boston, NH 03070
Phone: (603) 487-2955

Natural Power, which also makes solar equipment, manufactures a remarkable selection of anemometers and wind monitors in general. If there is *anything* you want to know about the wind, one of Natural Power's instruments will do it. They also manufacture control panels, dynamic loading switches and alternators for use with wind power, the 1.5 kw alternator costing $290 for 12 and 24 V models.

Natural Power anemometer/recorder.

Natural Power's Octahedron Module tower, self-supporting and based on a design by Windworks, is fully hot-dipped galvanized and can take a wind loading of up to 125 mph with a wind generator up to a 20-ft. rotor diameter. It can be erected from the ground up, thus eliminating the need for crane or gin pole.

25-ft. tower	$ 725
34-ft. tower	$ 925
42.5-ft. tower	$1,150
51.5-ft. tower	$1,450
70-ft. tower	$1,900
89-ft. tower	$2,600

Tower tops are $100, with bearings $200.

Natural Power control panel.

Natural Power Ltd.
Yorkshire House
Greek Street
Leeds LS1 5SX, England
Phone: (0532) 468146

This firm, not associated with the previous one, manufactures the Vortex Aero Generator, a twin-bladed upwind mill rated at 3.5 kw in a 22 mph wind. It costs about $3,500.

North Wind Power Company, Inc.,
Box 315, Warren, Vermont 05674
Phone: (0532) 468146

North Wind is best known for its re-built Jacobs Wind Electric plants. Jacobs manufactured and sold tens of thousands of wind generators in the period 1931-1956, and in that time they proved to be sturdy machines "built to last a lifetime." North Wind uses Jacobs generators as the base for their completely rebuilt machines. Each generator is reconditioned, the armature rebuilt, new brushes and bearings installed, and finally is tested. Reconditioned or new propellers are supplied with either fly-ball or blade-control governors.

Jacobs are driven by 3-bladed, 14-foot diameter propellers made from aircraft-quality Sitka spruce. The propeller direct-drives a special slow-speed 6 pole DC generator, which gives its full output at a mere 350 rpm in wind speeds of under 20 mph. Each generator weighs over 450 pounds, the field coils alone containing close to 75 pounds of copper.

The carbon brushes used with the Jacobs are known to last up to 20 years, a vast improvement on most brushes which rarely last more than one year. The direct-drive from propeller to generator avoids the cost and trouble of gears. Indeed, there are many who believe the Jacobs, especially the later 3 kw model, to be the best and most durable wind generator ever manufactured. Certainly one can say that these machines were never infected with the modern industrial disease of "built-in obsolescence."

Prices:	Jacobs 2 kw, 32 volt	$2,200
	Jacobs 2 kw, 110 volt	$3,500*
	Jacobs 3 kw, 32 volt	$3,200
	Jacobs 3 kw, 110 volt	$4,600*

* With new propellers and blade-control governors.

North Wind also sells new and used self-supporting and guyed towers. Their catalog/information package (well worth the $2 charged) also contains details of their extensive range of batteries and solid state and rotary inverters. North Wind also sells light bulbs and motors which operate on 32 volt DC.

Apart from dealing in wind power, North Wind provides a complete energy service, for planning the integration of wind, water, wood and solar energy in the best possible way.

See also Jacobs installation report, page 61.

"Admiral Richard E. Byrd took a Jacobs direct-drive wind generator to Antarctica in 1933. Its spruce wood propellers were still spinning when the explorer inspected the base 14 years later. Cost of maintenance — nothing."

National Geographic, December, 1975

Pinson Energy Corporation
Box 7
Marstons Mills, MA 02648
Phone: (617) 477-2913

Pinson Energy has just started to manufacture the "Cycloturbine,"
a self-starting variable pitch vertical-axis wind generator. Herman
Drees and his team of highly qualified helpers has been building
and testing the "Cycloturbine"
(see photo 22 under Windmills,
page 33) for years — as all windmills should be before they are
manufactured and sold. The blade
length is 8 ft. and the rotor diameter is 12 ft. It will start in a
5 mph wind and gives a maximum
of 4 kw at 30 mph (2 kw at 24
mph). The unit, complete with a
30-ft. tower costs $5,000.

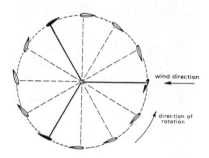

Operation of the PEC Cycloturbine.

P.I. Specialist Engineers Ltd.
The Dean
Alresford
Hants, England

This British firm is currently preparing to manufacture a variable-
pitch vertical-axis wind generator, see Figure 23 under "Windmills"
on page 34, which will have two wooden blades rotating on an
aluminum arm and will generate 500 watts in a 15 mph wind. The
turbine, based on the work of Dr. Peter Musgrove at Reading
University, will use an electric start since the rotor is not self-starting. The expected price is about $2,800.

Product Development Institute
508 S. Byrne Rd.
Toledo, OH 43609
Phone: (419) 382-3423

The 3 kw Wind Genni.

PDI has developed the Wind Genni shown at left, a 3-bladed upwind machine. Its all-fiberglass blades are something I would want to know a lot more about before purchasing. No details of maximum design wind speed are included in the company information packet. The machine starts charging at 9 mph wind and reaches a maximum output of 3,000 watts at 20 mph. The blades automatically feather in high wind speeds.

The Wind Genni is intended for use interfaced with the power company line, so that the "Base Load Injector" developed by PDI serves exactly the same function as the Gemini Inverter — it converts varying voltage output from the generator to standard grid voltage when connected to the grid. The mill, complete with the Base Load Injector is priced at a reasonable $3,595. The company sells 40-ft. towers for $575, 60-ft. for $700.

Sencenbaugh Wind Electric
2235 Old Middlefield Way
Mount View, CA 94306
Phone: (415) 964-1593

Sencenbaugh Wind Electric, started in 1972 as an agent for Dunlite (see above), has now developed its own range of products as follows:

Sencenbaugh Wind Generator Model 1000-14 (rated 1 kw at 22 mph wind). Upwind horizontal-axis machine with 12 ft. diameter 3-bladed propeller, constructed of machine-carved Sitka spruce with bonded copper leading edge. The propeller speed is 175 rpm at cut-in and 290 rpm at maximum output. Transmission is through a helical gear with a 3:1 ratio and is over-designed to use only 25 per-

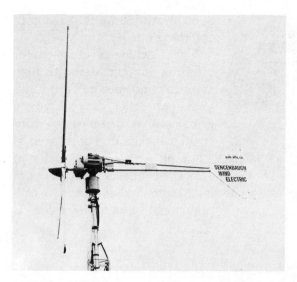

Sencenbaugh Model 1000-14: 1000 watts in a 22-mph wind.

cent of rated capacity at full output. Gear and alternator, a 3-phase 6-pole low-speed type, both are sealed in a cast aluminum body. Maximum continuous output at 14 volts DC is 1,000 watts at 22-23 mph; peak output is 1,200 watts (180 watts at 10 mph). The cut-in and charge rates are electronically controlled by a solid state speed sensor and voltage regulator pioneered by Sencenbaugh in 1973.

Propeller overspeed control is provided by utilizing the increase in wind pressure on the propeller and propeller thrust to furl the propeller downwind in high wind speeds. The foldable tail automatically begins to fold because of this force, in winds above 25-30 mph. Re-opening of the tail is automatic, being gravity-fed — a neat arrangement which provides a simple yet positive method of overspeed control. It also protects the windplant from gales and airborne debris. Maximum design wind speed limit is 80 mph and the turbine is built to a safety factor of 1.5. The price of the 12 or 24 V model is $2,650 and includes the control panel but not a stub tower.

For a small additional cost a special marine model is available with stainless steel hardware, highly recommended for coastal areas.

Sencenbaugh Model 500-14. Except for the following, the characteristics of this model are similar to the one above:

The 500-14 has an output of 500 watts at 25 mph, a peak output

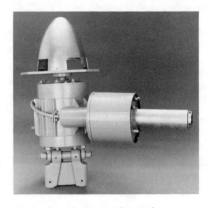

Sencenbaugh 500-14 direct-drive model.

of 600 watts (50 watts at 12 mph), and the propeller is only 6 ft.

This machine is designed for use in severe climates with high average wind speeds. Thus it has a small propeller, direct drive (no gears) and is designed to withstand a maximum wind speed of 120 mph with a safety factor of 1.5. The propeller speed at cut-in is 280 rpm, and at maximum output 1,000 rpm. The price is $2,200.

Output chart for Sencenbaugh Model 500-14.

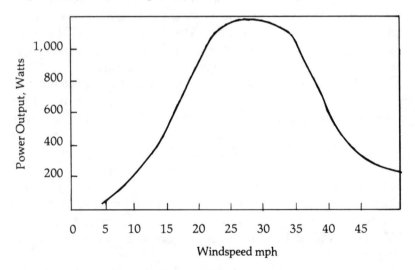

Sencenbaugh Model 24-14. This is a 24-watt wind generator designed to trickle charge 12-volt batteries on boats. The 20-inch-diameter propeller direct-drives a permanent magnet DC generator. The maximum output is rated at 21 mph. The price is $385.

Electronic Wind Odometer (or anemometer). This instrument records the average wind speed for a given time. The cup-contact transmitter responds to the wind and this response is electronically recorded on the separate odometer which can be situated up to 500 feet away and is battery operated. This enables one to ascertain the average wind speed where the transmitter is placed. The price is $115.

Sencenbaugh also supplies batteries, inverters, towers and a complete installation service. For further details send $2 for their catalog.

Solar Plexus
West Street
Hadley, MA 01035

Solar Plexus makes urethane polymer blades for the home windmill builder. Blades, each 5 ft. long, are strengthened with a metal spar running down the middle. Used in sets of three, they will produce 1.5 kw in a 20 mph wind with a tip speed ratio of 5:1. A set of 3 blades costs $92.50 shipped freight collect.

Topanga Power
Box 712
Topanga, CA 90290
Phone: (213) 455-2458

Topanga manufactures a home construction kit based on the Helion 12/16 plans (see page 93). The kit is sold in a semi-finished state or as individual components. Tasks left to the builder are final polishing and fitting of bushings, painting carriage and tower adaptor, bolting components into place, and pop-riveting the blades together. This work takes about two man days. (See also *Kedco, Inc.* for similar but fully manufactured and assembled model.)

The resulting 3-bladed downwind machine can be supplied with a 12- or 16-ft. diameter propeller. Due to alternator limitations, the maximum power output will be the same with either set of blades, but the advantage of the longer blades is that full power will come at a much lower windspeed. The long blades deliver full power at 18 mph, and the short blades at 21 to 22 mph. Maximum power from the 12 V alternator is 1,200 watts and from the 24 V model is 1,800 watts.

The blades are made from aluminum and transmission is via a gearbox with an 8.6:1 ratio. The weighted, self-feathering blades will continue to develop full power even when partially feathered. Ground-operated shut off is available as an optional extra. The price is $1,400, excluding stub tower.

TWR Enterprises
355 South Riverside
Rialto, CA 92376

TWR sells a range of plans and home-built wind generator kits. Send $1 for details.

Unarco-Rohn
6718 West Plank Road
P.O. Box 2000
Peoria, IL 61601
Phone: (309) 697-4400

For several years Rohn has supplied both self-supporting and guyed tower for use with Elektro, Dunlite, Jacobs, Sencenbaugh, Aero Power, and many other wind generators. Rohn towers also come complete with top sections suitable for each type of wind generator.

Guyed tower suitable for Sencenbaugh mills (with top section)

40-ft. tower	$500	80-ft. tower	$865
60-ft. tower	$600		

Guyed tower suitable for Dunlite & Elektro (with top section):

40-ft. tower	$ 600 (approx)	80-ft. tower	$1,255 (approx)
60-ft. tower	$ 925 (approx)	100-ft. tower	$1,600 (approx)

Self-supporting tower with top section for Dunlite & Elektro:

| 40-ft. tower | $1,150 | 80-ft. tower | $2,200 |
| 60-ft. tower | $1,600 | 100-ft. tower | $2,925 |

The Wind Energy Supply Co. Ltd.
Bolney Avenue
Peacehaven
Sussex BN9 8HQ, England

WESCO is the combination of two advanced technologies, one specializing in helicopter rotors and the other in control systems. Together they plan to manufacture a full range of wind machines adapted to various uses.

The gaunt machine shown (see photo below) is the Oleo Hydraulic, used for direct heating by oil, without electronics at all. The 60-ft. propeller is supported by a 45-ft. steel tower. The two downwind blades are made of fiberglass with steel spars, the basic design based on many years of experience with helicopter rotors.

WESCO's 60-foot-diameter Oleo Hydraulic windmill.

The variable-pitch rotor will produce about 7 kw at 10 mph wind-speeds and a maximum of 190 kw at 30 mph. On a site with an annual average wind speed of 11 mph the output of heat energy is likely to be about 150,000 kwh a year. Weighted flaps on both blades prevent the propeller from exceeding a maximum of 120 rpm.

The propeller shaft and initial gear-up are made of a rear half-shaft from one of those huge earth-moving trucks. The mechanical energy from the shaft is then converted to heat energy by hydraulics. The hydraulic oil is then pumped from the tower head, through insulated pipes, to a heat exchanger in the greenhouse. There, the heat is transferred to water circulating through the greenhouse heating system.

Even though WESCO is happy with the results of its initial tests, I remain a little wary of twin-bladed downwind mills because of the apparent failure of the similar ERDA 100 kw mill. The ERDA mill suffered from excessive vibration as the blades passed through the "wind shadow" cast by the tower. The next few months of tests will show if my doubts are ill-placed or not.

WESCO plans to manufacture a complete range of wind machines with outputs ranging from 0.5 to 200 kw of electricity or heat equivalent. The propeller diameters in their basic range will be 10, 16, 23 and 60 feet. Apart from their novel hydraulic, direct-heat system, WESCO also is involved in developing wind systems in conjunction with chemical heat storage, solar collectors and heat pumps.

WESCO has at least two other mills on their drawing boards. One is a three-bladed upwind mill sold as a home-assembly kit and used for domestic wind-electric systems. The second is a remarkable wind-electric system driven by a single propeller blade. Yes — half a propeller and self-starting, too! The idea is that with such a low blade solidity high speed is attained, resulting in a wind generator giving its maximum output of 500 watts in an 11 mph windspeed!

To complement their range of wind generators WESCO has developed an interesting electronic control system (patents pending) which in operation will pass current, at a pre-selected voltage and frequency, direct to the electric load, thus by-passing the batteries and therefore reducing the quantity of expensive batteries required. When the output from the wind generator exceeds the load, the excess is fed to batteries, which in turn supply power when output is less than load. The enormous benefit of this electronic control system became obvious when used with a wind generator giving its maximum output in an 11 mph wind speed. For data on a WESCO greenhouse-heating installation, see page 76.

Winco
Dyna Technology, Inc.
East Seventh at Division Street
P.O. Box 3263
Sioux City, IO 51102
Phone: (712) 252-1821

The Winco Wincharger 200 watt has been around for a long time, (see also earlier under Coulson Wind Electric). The 6-ft. propeller direct-drives a fairly sturdy 4-pole generator, though both generator and slip ring brushes do need occasional replacing. The mill is available in 12, 24, 28, 32, & 36 volts and comes complete with instrument panel and 10-ft. stub tower for about $450. Maximum output of 200 watts is at 23 mph.

Winco Wincharger 200 watt.

Average Usable kwh per Month

10 mph average	20
12 mph average	26
14 mph average	30

For more data on Wincos, see pages 53, 58-60.

Wind Power Systems, Inc.,
8871-A Balboa Avenue
San Diego, CA 92123
Phone: (714) 560-9452

Wind Power System's RD-4000.

Wind Power Systems (WPS) has developed a unique wind generator operated by three propellers, the rims of which drive the alternator. The RD-4000, not yet commercially available, gives a maximum output of 4 kw in a 20 mph wind, and the cut-in speed is 7 mph. Output from the 3-phase brushless alternator is 110 volts DC. To protect the rotor in high winds, all three propellers tilt upwards, and are hydraulically returned in normal wind speeds.

WPS has designed and built a Dutch four-arm mill which helps to power a motel-restaurant complex at Santa Nella, California. The 46-foot diameter propeller drives an alternator which generates about 8 kw in 20-mph wind speeds. The windmill, first of its kind to be built for a long time, is capable of withstanding winds of up to 140 mph!

WPS also manufactures a "Windplant Energy Simulator" that, using an anemometer, simulates the performance of a wind generator on any site where the simulator is installed. The cost is $690.

Windworks
Box 329, Route 3
Mukwonago, WI 53149
Phone: (414) 363-4408

As an engineering consulting firm active in wind energy, power conditioning and load management, advanced structural systems and publishing, perhaps Windworks' most interesting line is the Gemini synchronous inverter, a solid state device which when interposed between a variable voltage DC power source and an AC public power grid, converts the DC to AC at standard line voltage and frequency.

In operation, all available DC power is converted to AC. If more power is available from the DC source than is required by the load, the excess flows into the power grid. If less power is available than is required by the load, the difference is provided by the power company in the normal fashion. For any variable and intermittent power source that is small in comparison to the capacity of the power line grid to which it is tied, use of such a synchronous inverter eliminates the cost of a storage battery system.

Compared to a wind system using a lead-acid battery bank for storage and a conventional inverter to supply all or part of the load, one using a synchronous inverter operates at a higher overall efficiency, thus increasing the useable energy. Wind systems using the Gemini are able to use or "store" all power generated. The power grid in most parts of the country may be used as a sink even for high-output wind generators. (For details on one such application, see page 61).

For wind systems with capacities up to 20 kw, synchronous inverters can be one sixth the cost of conventional inverters per kilowatt capacity and may result in up to a 50 percent reduction in the capital cost of the total system. With a higher system-efficiency and a lower capital cost for the same generating capacity, the kilowatt-hour cost is considerably less than with battery/inverter systems. Because the nature of the service here provided by the public utility — that of a storage medium and current regulator — differs from their usual role, it will be important to establish some initial agreement with them. The singular disadvantage of the Gemini is this dependence upon the public power grid, but even so, a battery/inverter can be installed if or when desired.

Windworks can supply Gemini inverters rated to any capacity. The conversion efficiency at maximum power is 95 percent. The following gives an example of the costs:

> 4 kw maximum conversion power $ 780
> 8 kw maximum conversion power $1,450

(For more on the Gemini Inverter, refer back to pages 49 and 61).

Windworks developed the octahedron tower now manufactured by Natural Power, Inc. (see page 98). They publish *The 25 ft. Sail Windmill Plans*, reviewed in the *Bibliography*, and also a *Wind Energy Chart* ($3.25), which is a fine visual depiction of the chronology of wind power development, with general wind power data carried on the reverse. Windworks' *Wind Energy Bibliography* is a good buy at $3.

WTG Energy Systems, Inc.
P.O. Box 87
One La Salle Street
Angola, NY 14006
Phone: (716) 549-5544

WTG has just completed the erection of a 200 kw wind generator, using a 3-bladed propeller with a diameter of 80 feet. The design is based on the Dutch Gedser mill, the most successful large wind generator ever built.

The new mill will supply power to the residents of Cuttyhunk Island, Massachusetts, and it will be interesting to see how this machine compares with the ERDA 100 kw wind generator.

Zephyr Wind Dynamo Co.
P.O. Box 241
Brunswick, ME 04011
Phone: (207) 725-6534

The Zephyr Wind Dynamo is a three-bladed downwind machine with a maximum output of 15 kw at 30 mph (500 watts at 10 mph). The low-speed, direct-drive alternator is specifically designed for wind-driven use and is of the permanent magnet type. The blades are a lightweight composite of an injection-molded urethane foam

Zephyr Wind Dynamo.

body and coated Kevlar skin. Overspeed protection is by glide-out spoilers located near the blade tips. The maximum tip speed ratio is 6.5:1. (See page 71 for installation example.)

The price of the Zephyr, including control panel and 14-foot tower mast, is approximately $12,000. This figure varies according to the intended use of the machine. While Zephyr has sold three of their machines, two for test purposes, they still regard it as experimental. Should the tests and time prove the design to be a good one, then confidence will increase, as will production, thus causing a reduction in cost.

Zephyr has also built a prototype of a rather unusual-looking slant-axis, wind machine called the *Tetra-helix*. The drawing outlines the idea. The whole structure is rigid under tension and will collapse easily without tension. Zephyr is currently testing a 2.5 kw prototype and the results should be interesting.

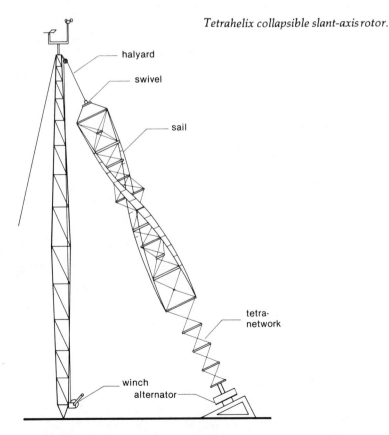

Tetrahelix collapsible slant-axis rotor.

halyard

swivel

sail

tetra-network

winch
alternator

Zephyr also manufactures a smaller model called the *"Tetra-helix-S,"* mainly intended for trickle charging batteries on small boats. The sails made of Dacron sailcloth, describe a helical pattern about the torque-delivery system, resulting in a self-starting, omni-directional rotor. Maximum output from the permanent-magnet alternator is 7 watts in a 25 mph wind.

MECHANICALLY
APPLIED WINDMILL
MANUFACTURERS

Water Pumping

Aermotor
Industrial Park, P.O. Box 1364
Conway, AK 72032

KMP Parish Windmills
Box 441
Earth, TX 79031

Dempster Industries, Inc.
P.O. Box 848
Beatrice, NB 68310

Merritt Windmill, Inc.
P.O. Box 1374
Merritt, B.C.
Canada VOK 2BO

O'Brock Windmill Sales
Route 1, 12th Street
North Benton, OH 44449

Hydraulic Oil Direct Heating

The Wind Energy Supply Co. Ltd..
Bolney Ave.
Peacehaven
Sussex BN9 8HQ England
See page 107 for details.

AGENTS

Alaska Wind & Water Power
P.O. Box 6
Chugiak, Alaska 99567
Phone: (907) 688-2896

Active agents for a complete range of wind generators and water turbines.

Alternate Energy Systems
150 Sandwich Street
Plymouth, MA 02360
Phone: (617) 747-0771

Automatic Power
P.O. Box 18738
Houston, TX 77023
Phone: (713) 228-5208

The Big Outdoors People
2201 N.E.Kennedy Street
Minneapolis, MN 55413
Phone: (612) 331-5430

Main activity is in geodesic dome housing, but also involved in alternative energy and design of wind power system.

Boston Wind
2 Maston Court
Charlestown, MA 02129

Boston Wind sells new and used wind generators and accessories and offers an installation service. It also holds courses, workshops and slide-lectures and publishes a quarterly newsletter.

Budgen & Associates
72 Broadview Avenue
Pointe Claire, P.Q. H9R 3Z4
Canada
Phone: 731-3431

Dr. Harry Budgen is technical advisor on alternative energy to Brace Research Institute of Macdonald College, which has a Lubing, a 5 kw Elektro and a 25 ft.-diamater sail windmill on campus. Budgen & Associates supply Lubing and Dunlite wind generators and pumps and also plans for a Brace-designed wind generator with three-bladed, 32-ft. diameter propeller, developing 49.5 hp in 30 mph winds.

Clean Energy Systems
RD 1, Box 366
Elysburg, PA 17824
Phone: (717) 799-0008

Offers mechanical engineering design assistance on wind power and other renewable energy sources.

Earthmind
5246 Boyer Road
Mairposa, CA 95338

This group of wind power enthusiasts includes Michael Hackleman, author of two excellent books: *Wind & Windspinners* and *The*

Homebuilt, Wind-Generated Electricity Handbook, (see *Bibliography*). Earthmind deals mainly with reconditioned Jacobs and Winchargers, and since it is a non-profit organization, proceeds from sales go towards establishing its research center.

Energy Alternatives, Inc.
69 Amherst Road
Leverett, MA 01054
Phone: (413) 549-3644

These agents for Dunlite, Jacobs, Elektro and Aero Power windmills also offer integrated sun, wind and wood designs.

Enertech Corporation
P.O. Box 420
Norwich, VT 05055
Phone: (802) 649-1145

Enertech, a company with years of practical experience in wind power, is agent for Sencenbaugh, Dunlite, Elektro and Winco. Henry Clews and his Solar Wind Company have joined forces with Enertech, and they sell everything required for a wind installation — from anemometers to inverters — and also are agents for the inexpensive Dutch Sparco wind pump. Send $2 for their booklet, "Planning a Wind-powered Generating System."

Environmental Energies, Inc.
P.O. Box 73
Front Street
Copemish, MI 49625
Phone: (616) 378-2000

Founded by Al O'Shea, one of the organizers of the American Wind Energy Association (see page 123), EEI deals in Dunlite, Winco, rebuilt Jacobs and Elektro machines. It offers a complete installation service and stocks a line of Creative Electronics inverters. EEI prac-

tices what it preaches: The shop is powered by an Elektro and is
heated by a wood-burning stove and solar energy. Send $5 for their
detailed and informative booklet on wind, solar and other energy
systems.

Environment Systems Ltd.
5707 Indian River Road
North Vancouver, B.C. V6G IL3
Canada

I have no current information on this company.

Future Resources & Energy Ltd.
167 Denison Street
Markham, Ontario L3R IB5
Canada
Phone: (416) 495-0720

Fred Drucker of FRE, during his frequent visits to Elektro in Switzerland over the past three years, has developed a thorough understanding of Elektro wind generators. As a result FRE is now the sole
Canadian agent for Elektro. Each machine is checked for quality
control before erection. FRE specializes in wind and solar combinations.

Independent Energy Systems
6043 Sterrettania Road
Fair View, PA 16415

IES offers reconditioned Jacobs. Send $2 for information on wind,
solar and wood services.

Independent Power Developers
Box 1467
Noxon, MT 59853
Phone: (406) 847-2315

IPD, agent for Dunlite machines, was recently awarded a contract by the Montana Department of Natural Resources to build and demonstrate a 15-ft. diameter, three-bladed downwind machine to develop 18 kw in a 33 mph wind. The blades will be made of aluminum.

Jopp Electrical Works
Princeton, MN 55371

Martin Jopp, rightly referred to as a "wind wizard," started back in 1917, when he built a few hundred Jopp Wind generators. His own home has been powered by two 3 kw Jacobs wind generators for years. Jopp now turns out new parts for old Jacobs mills at very reasonable prices. What he doesn't know about Jacobs is not worth knowing. He writes on the subject, answering letters in *Alternative Sources of Energy*.

Life Size Aero Design
432 Franklin Street
Alburtis, PA 18011

Data on the activities of this and the following listed companies are lacking.

Natural Energy Systems, Inc.
55-A West Avenue
Box 491
Wayne, PA 19087
Phone: (215) 293-0116

Prarie Sun & Wind Company
4408 62nd Street
Lubbock, TX 79409

Agent for Winco & Aero Power.

Quirks
33 Fairweather Street
Bellevue Hill
NSW 2023, Australia

Markets the Dunlite wind generator.

Real Gas & Electricity Co.
P.O. Box 193
Shingletown, CA 96088
Phones: (916) 474-3456
(707) 526-3400 (Santa Rosa office)

Designs and installs systems using Dunlite and Elektro wind generators. Offers complete installation service or supervisory assistance. They also deal with solar and water power.

Rede Corporation
P.O. Box 212
Providence, RI 02901
Phone: (401) 861-5390

Rede is U.S. agent for DAF Darrieus Wind generators.

Sigma Engineering
Box 5285
Lubbock, TX 79417
Phone: (806) 762-5690

Regional distributor for DAF Darrieus rotors. See Rede Corporation above.

Solar Energy Co.
810 18th Street
Washington, D.C. 20006

No data on this company is available.

Sunstructures, Inc.
Integrated Architectural Design
201 E. Liberty Street, No. 6
Ann Arbor, MI 48104
Phone: (313) 994-5650

Specializes in integrated wind and solar systems. Also conducts workshops.

Total Environmental Action
Church Hill
Harrisville, NH 03450
Phone: (603) 827-3374

Offers a wide range of architectural and engineering services in the alternative energy field. Emphasis on research, design and teaching.

Wind Energy Systems (Sunflower Power Co.)
Route 1, Box 93-A
Oskaloosa, KS 66066
Phone: (913) 597-5603

WES sells second-hand wind generators — Jacobs, Wincharger, Wind King, etc. — is agent for the Gemini inverter, and designs and builds integrated energy systems. Manager Steve Blake, who has also had extensive experience in building and testing Savonius rotors at the Brace Research Institute (see Budgen & Associates, page 117), also works with the Appropriate Technology Group at the same address.

Windependence Electric
P.O. Box M 1188
Ann Arbor, MI 48106
Phone: (313) 769-8469

Windependence sells a range of reconditioned wind generators — Allied, Winpower, Jacobs, etc.

BIBLIOGRAPHY

The most important and up-to-date writing on wind energy available in the United States is contained in the *Wind Power Digest* (see under listing).

Alternative Sources of Energy

A very fine journal, many excellent articles on wind power, inc. "Martin Answers" by Martin Jopp. ASE is *the* journal for the home builder of windmills. It contains lots of solid, safe and intelligent advice. They have published many plans (and by far the best) for home windmill builders. ASE No. 24 (Feb '77) is a special wind power issue. Six issues yearly are well worth the $10 (foreign $16) from: Route 2, Milaca, MN 56353.

How to Construct a Cheap Wind Machine for Pumping Water, A. Bodek. $1.25

Performance Test of a Savonious Rotor, Simonds & Bodek. $2

Notes on the Development of the Brace Airscrew Windmill as a Prime Mover, R. Chilcott. $.50

A Simple Electric Transmission System for a Free-Running Windmill, Barton & Repole. $2

These publications available from Brace Research Institute, MacDonald College, Ste. Anne de Bellevue, Quebec HOA 1CO, Canada.

American Wind Energy Association
Box 329   Route 3,
Mukwonago, WI 53149

The AWEA is a national association that represents manufacturers, distributors, and researchers involved in the development of wind energy.

Membership is $25 per year. The AWEA issues a quarterly newsletter, publishes the *Wind Technology Journal* and holds convivial and enlightening conferences.

Catch the Wind, Landt & Lisl Dennis

A well-written general introduction to wind power. $7.95 from Four Winds Press, 50 West 44th Street, New York, NY 10036.

Do-it-Yourself Sail Windmill (Cretan) Plans

How to build the 12-foot diameter Sail Wind Generator, 200 watt output at 15 mph, maximum output 300 watts, shown in photograph on page 57. Cost: $2 from National Center for Alternative Technology, Machynlleth, Powys, Wales, England.

Electric Power from the Wind, Henry Clews.

This booklet describes the systems used with a 6 kw Elektro and a 2 kw Dunlite at the author's home, also makes interesting comparisons between the two. Costs $2 from Enertech, P.O. Box 420, Norwich, VT 05055.

Energy from the Wind, Burke & Meroney

The bibliography of all bibliographies on wind energy — complete and annotated book, costs $7.50. Also available, a "First Supplement" (from 1975 to 1977), cost $10. Both available from Publications, Eng. Research Center, Foothills Campus, Colorado State University, Fort Collins, CO 80523.

Food from Windmills, Peter Frankael

Describes the windmill-building activities of the American Presbyterian Mission in Ethiopia. Contains lots of good details on how to build sail mills — mainly 11-foot-diameter and used for water pumping. Available for $6.00 (f2.90) from Intermediate Technology Publications, 9 King Street, London WC2, England.

The Generation of Electricity by Wind Power, E.W. Golding

Originally published in 1955, this remains a technical masterpiece on wind power. From Halstead Press, New York or Spon Ltd., London, England.

Helion Model 12/16 Windmill Plans

See Helion entry under *Manufacturers* for details.

The Homebuilt, Wind-Generated Electricity Handbook, Michael Hackleman

That this book is mis-titled does not detract from its brilliance as a complete guide to *re-building* and installing old Jacobs & Wincharger machines. Any who have or intend to buy such machines (or any old wind generator) are well advised to buy this book at $8 from Earthmind, 5246 Boyer Road, Mariposa, CA 95338.

Homemade Windmills of Nebraska, Erwin Barbour

This reprint, originally published in 1898, describes how to build "weird & wonderful" windmills for pumping and sawing. Available from Farallones Institute, 15290 Coleman Valley Road, Occidental, CA 95465.

The Journal of the New Alchemists

Journal No. 2 contains details of how to build a Sailwing rotor. Available for $6 from N.A.I., P.O. Box 432, Woods Hole, MA 02543.

The Mother Earth News
P.O. Box 70
Hendersonville, NC 28739

Mother Earth News, the magazine for self-sufficiency, carries occasional articles on wind power, but we could do with more. Well worth $10 for 6 issues a year. TMEN also reprinted *Homemade Six-Volt Wind-Electric Plants*, originally published in 1939. Send $1.

Shades of Rube Goldberg:

"As you can see, the automatic shut-off device for the Gedser mill is a pipe that comes up from the floor, and on top of the pipe is a cup in which sits a heavy oversized ball. That ball is connected by a string to an old-fashioned Square D type switch on the wall. When the tower starts to vibrate, the ball rolls out of the cup and the string pulls the switch to stop the machine."

The Princeton Sailwing Program, Dr. T. Sweeney

A short report on the two-bladed Sailwing developed at Princeton University. Send $2 to Forrestal Campus Library, Princeton University, P.O. Box 710, Princeton, NJ 08540.

Rain
2270 N.W. Irving
Portland, OR 97210

A fine monthly "Journal of Appropriate Technology." April '77 issue contained an excellent article on the 200 kw Dutch Gedser mill — a mill much praised for its low cost and suitability for local manufacture. A *Rain* subscription is $10, a single issue $1.

Reinforced Brickwork Windmill Tower, A.B. Bird

Mainly on designing and building brickwork towers, this book also contains a section on how to build the 40-foot diameter Sail Windmill photographed on page 22. Costs $3 (£1.15) from Structural Clay Products Ltd., 230 High Street, Potters Bar, Herts, England.

Simplified Wind Power Systems for Experiments, Jack Park

The core of this book contains data of interest to home builders on the aerodynamic, structural and mechanical design of wind-driven propellers. $8 from J. Park, Box 4301, Sylmar, CA 91342.

Wind Power Digest
54468 CR31
Bristol, IN 46507
Phone: (219) 848-4360

WPD is essential and most enjoyable reading for the potential wind generator owner. Mike Evans is the editor, Jack Park the technical editor, and Joe Carter does some brilliant reporting on and interviews with prominent wind power people. All in all the *Digest* successfully reflects the excitement and enthusiasm of the wind power movement. At a yearly (four-issue) subscription rate of $6 (back issues $2), you can't go wrong.

Wind Energy Bibliography, from Windworks

A compilation of books, articles and papers with sections that include wind, windmills, aerodynamics, electrics, towers, storage, conversion, hydrogen and commercial units. As much of a bibliography as one would need, well worth the $3 cost. Windworks, Box 329, Rte. 3, Mukwonago, WI 53149.

Windustries

A fine regional quarterly newsletter, mainly on wind power; subscription, $10 yearly ($15 institutions). Available from Great Plains Windustries, Box 126, Lawrence, KS 66044.

Wind and Windspinners, Michael Hackleman

The first half of this book covers the electrics of wind systems, the second tells how to build Savonius rotors. Definitely *the* book on the S-rotor. From Earthmind, 5246 Boyer Road, Mariposa, CA 95338.

Windy Ten Dutch Windmill Plans

For those who want to build a Dutch Four-arm. Propeller is 8-ft. diameter, can produce up to 500 watts if coupled to a generator. Suitable for the serious romantic only. Plans $16 from Edmund Scientific, 1006 Edscorp Building, Barrington, NJ 08007.

25-Foot Diameter Sail Windmill

Detailed design manual for a 6-sail mill (and a 42-foot octahedron module tower with a platform for sail reefing), intended for mechanical purposes or for gearing up to generate electricity. Requires manual reefing in winds over 20 mph. Typical power outputs are as follows:

Wind Speed	Power Available
5 mph	0.1 hp
10 mph	0.8 hp
15 mph	2.8 hp
20 mph	6.7 hp

Plans, originally prepared for Brace Research Institute by Windworks, available for $25 from Windworks, Box 329, Route 3, Mukwonago, WI 53149.

The following four books all include articles on wind, water, solar, biofuels and integrated energy designs:

New Low-Cost Sources of Energy for the Home, Peter Clegg, Garden Way Publishing. 1976 update.

Energy Primer, Portola Institute.   1977 update.

Other Homes & Garbage, Lockie, Masters, Whitehouse & Young, Sierra Club Books. 1975.

Producing Your Own Power, edited by C. Stoner, Rodale Press. 1975.

INDEX

Other Garden Way Books
You Will Enjoy

The owner/builder and the home-owner concerned about energy conservation and alternate construction methods will find an up-to-date library essential. Here are some excellent books in these areas from the publisher of *Harnessing The Wind for Home Energy*.

Harnessing Water Power for Home Energy, by Dermot McGuigan. 112 pp., quality paperback, $4.95. An authoritative, detailed look at the uses of water power for small-scale operations.

Low-Cost Pole Building Construction, by Douglas Merrilees and Evelyn Loveday. 118 pp., deluxe paperback, $5.95. This will save you money, labor, time and materials.

Build Your Own Stone House, by Karl and Sue Schwenke. 156 pp., quality paperback, $5.95; hardback, $8.95. With their help, you can build your own beautiful stone home.

New Low-Cost Sources of Energy for the Home, by Peter Clegg. 250 pp., quality paperback, $7.95; hardback, $12.95. Covers solar heating and cooling, wind and water power, wood heat and methane digestion. Packed with information.

Wood Stove Know-how, by Peter Coleman. 24 pp., illustrated paperback, $1.50. Installation, cleaning and maintenance instructions, plus much more.

The Complete Book of Heating with Wood, by Larry Gay. 128 pp., quality paperback, $3.95. Fight rising home heating costs and still keep very warm.

The Complete Homesteading Book, by David Robinson, 256 pp., quality paperback, $5.95; hardback, $12.95. How to live a simpler, more self-sufficient life.

Buying Country Property, by Herb Moral. 128 pp., quality paperback, $3.95. Sure to be your "best friend" when considering country property.

Build Your Own Log Home, by Roger Hard. 204 pp., quality paperback, $6.95; hardback, $12.50. A real guidebook to building with logs or log home kits, by a man who has built his own log home and others.

Designing & Building a Solar House, by Donald Watson. 288 pp., quality paperback, $8.95; hardback, $12.95. The best and most thorough book yet on solar houses, with over 400 illustrations.

Your Energy-Efficient House, by Anthony Adams. 128 pp., quality paperback, $4.95. The perfect idea book for those concerned about saving energy in a new or existing house.

Building & Using Our Sun-Heated Greenhouse, by Helen & Scott Nearing. 156 pp., quality paperback, $6.95; hardback, $11.95. The Nearings share, in text and photographs, the secrets they have learned in over 50 years of gardening in New England all year-round.

Methanol & Other Ways Around the Gas Pump, by John Ware Lincoln. 144 pages; quality paperback, $4.95. How to "drive without gas"— using methanol—and a look at the past experiments and future politics of our gasoline supply.

These Garden Way books are available at your bookstore, or may be ordered directly from Garden Way Publishing, Dept. 171X, Charlotte, Vermont 05445. If your order is less than $10, please add 60¢ postage and handling.